W9-CEN-064

LURE OF THE INTEGERS

JOE ROBERTS

JOE ROBERTS

LURE OF THE

THE MATHEMATICAL
ASSOCIATION OF
AMERICA

NTEGERS

SPECTRUM SERIES

Published by
THE MATHEMATICAL ASSOCIATION OF AMERICA

Library of Congress Catalog Card Number 91-62053

ISBN 0-88385-502-X

Printed in the United States of America

Current Printing (last digit):
10 9 8 7 6 5 4 3 2 1

Preface

For a long time I have collected, in a rather haphazard way, interesting properties of various integers. If it came to my attention that, for example, 17 is the largest integer not the sum of three pairwise relatively prime integers each larger than 1, then this would be recorded on a slip of paper, along with its source, and put in my desk. I imagined it to be an entry in a mythical book titled *The Book of Integers.* On page n of this book would appear all of the interesting properties of the integer n.

Recently I have tried to bring some order into this collection of slips of paper and to fill in related details. As things progressed, the properties of the integers often led to other interesting and related bits of mathematics or to what seemed to be interesting sidelights to the properties themselves. Thus there were digressions and even digressions to the digressions. The result might, at first, appear to be a hodgepodge of facts and observations. It *is* that, but it also contains a thread of what a life devoted to mathematics might be like.

The sections of the book correspond, in order of appearance, to the positive integers discussed. Thus the section labeled Integer 17 will have to do with the integer 17 and related matters.

The book may be read consecutively from beginning to end or can be used for browsing purposes. I hope that dipping into it at any point will yield something of interest and, perhaps even for the professional, something new. It is perhaps regrettable, though not entirely so, that "recreational mathematics" is juxtaposed to "serious mathematics" and that only a professional in each subject at hand would know which is regarded as which.

A reader used to reading with paper and pencil at hand and proving all unproved assertions will have a hard time of it since it will not always be clear when this is a reasonable thing to do and when it is not. The text will not make this clear, and mathematics itself is notorious for not making this clear. On the other hand, there are many places where the reader will be expected to do just this sort of thing.

All in all the book is designed, first and foremost, to be interesting to anyone interested in numbers. I have tried to be fairly complete in providing enough references so that anyone wishing to trace what has been done will have a good entrée to the literature.

Finally, I would like to add a few comments reflecting something of my philosophical biases, thus making explicit an implicit underlying theme of the book which might otherwise only be apparent to one who read through the book from beginning to end. One might well conceive of these remarks as a justification for writing the book.

Mathematics is nonlinear in the sense that normally a mathematician working on a problem does not proceed in a straight path toward a solution. Often the resulting published work will concern something not even remotely considered at the time the investigation began. Even the day-to-day research work is often directed by chance: by what was done rather than by what was attempted, by the content of last week's colloquium, by a chance remark by a colleague or a student. The line demarking the separation of recreational from serious mathematics is not as clear as we may sometimes be led to believe.

Aside from the philosophical view that considers all mathematical activity as recreational, I believe that the preponderant amount of mathematical activity might actually be called recreational by those practitioners most successful in the world of serious mathematics. They would think that the activity deals with problems that are too specialized, with problems having insufficient application to other areas of science or to the mainstream of mathematics itself, with problems not making use of the more difficult and abstruse sides of the subject, with recreational problems, with subjects that are out of fashion.

None of these objections is necessarily relevant from the standpoint of whether or not the activity is suitable as the stuff of life for an individual mathematician. In any event, the present book is a sort of chronicle (encyclopedia?, handbook of interesting mathematics?, bibliography of contemporary activity?) of things, many of which fall outside of the mainstream of mathematics. Some of these things have been of interest to many people for an enormous number of years. Not a few of them come from recreational mathematics and some involve the most difficult realms of serious mathematics. Even a professional mathematician is not likely to be familiar with all of them.

I hope that something of the flavor, and interest, of what goes on in a mathematical community is apparent and that some of the inability of such a community to distinguish the difficult from the easy, the worthwhile from the non-worthwhile, the merely interesting from the merely useful, will be clear and understandable.

That seemingly simple problems will span the centuries is not uncommon. That these same problems have no current applications outside of their own intrinsic interest is not uncommon. That such problems may later find themselves thrust into the mainstream through other developments not related in any clear a priori fashion to the problem as originally investigated is not uncommon. Pure curiosity, rather than ultimate use of the results, has been one of the more dominant features in the investigations themselves.

None of this, in any way, indicates an opposition to applications; to the desire for coherence of large theories; to a conjoining of intellectual activity leading to unification. I *am* interested in excavations of the activities of generations of ordinary mathematicians—some in the mainstream, most of them not in the mainstream, but all of them captured by the fascination of numbers.[1]

---

[1] Though this book is not a book of numerical records it is, perhaps, worth mentioning that in the two years since the manuscript was completed some of the records mentioned no longer hold. As a single instance, note that on page 82 I imply that the largest known perfect number has 130,100 digits. Recently a 19-hour computation on a Cray-2 supercomputer found the 455,663-digit perfect number $2^{756838}(2^{756839} - 1)$. (See, e.g., *Focus* 12(1992) 3.)
—JBR, July 1992

Contents

Integers 2, 3

1 Equidecomposability

If one cuts a polygon into pieces with straight line cuts and reassembles the pieces into the same or another polygon, it seems clear the original polygon and the resulting polygon have the same area. Not only does this seem clear but it is, in fact, true.

The converse statement that, given two polygons with the same area, one of them may be cut with straight line cuts into a finite number of pieces which may then be reassembled into the other seems less obvious. Nevertheless this, also, is true and the assertion of its truth goes under the name of the *Bolyai–Gerwin theorem*.

Similarly, in 3 dimensions, if one cuts a polyhedron into a finite number of pieces by planar cuts and reassembles the pieces into the same or another polyhedron, then the original polyhedron and the resulting polyhedron have the same volume.

One might think, in analogy to the 2-dimensional case, especially after learning of the Bolyai–Gerwin theorem, that given two polyhedra of the same volume one of them may be cut into a finite number of pieces by planar cuts and the pieces reassembled into the other.

In his famous list of 23 problems (see Integer 23, Hilbert's List) Hilbert asked, in his third problem, whether or not it is true that two polyhedra of the same volume were "equidecomposable" in this way. As Novikoff says in the Foreword to *Hilbert's Third Problem* by Boltianskii

> The problem is not a perverse exercise in ingenuity; on the answer depends whether or not there exists an "elementary" theory of planimetry, which exploits only the area formula for rectangles, but makes no appeal to limit processes such as Archimedean "exhaustion" or Cavalieri's principle.

This problem was one of the very first of Hilbert's problems to be solved. Two years after its proposal M. Dehn proved that a cube and a rectangular tetrahedron of the same volume are not so related; i.e., one may not cut one of them into a finite number of pieces by planar cuts and reassemble the pieces into the other.

Complete expositions of these results may be found in the books cited below. (The first two books listed are translations from the Russian and have the same author. However, transliteration being what it is, the author's name is given two different spellings. In each case we keep the spelling found in the published translation.)

See also Integer 5, The Banach–Tarski Paradox.

REFERENCES

[1] V. G. Boltianskii, *Hilbert's Third Problem,* Wiley, New York, 1978, p. ix.

[2] V. G. Boltyanskii, *Equivalent and Equidecomposable Figures,* D. C. Heath, Boston, 1963.

[3] H. Eves, *A Survey of Geometry,* Chapter 5, Allyn and Bacon, Boston, 1972.

2 Cubing Cubes

As is discussed in Integer 21, Squared Squares, a square may be completely filled with a finite number of smaller squares no two the same size. Knowing this, one might enquire if it is possible to completely fill a cube with a finite number of smaller cubes no two the same size.

The answer is that it is not possible to do this. This answer is not only somewhat surprising in itself, but also in the fact that it is not at all difficult to prove. We sketch the proof in the next paragraph.

The proof is by contradiction. First, we assume that a cube may be completely filled with a finite number of smaller cubes of different sizes. In this cubing, the cubes, sitting on the bottom of the filled cube, induce on the bottom square of the filled cube a squaring of this square. (I.e., an exact splitting of this bottom square into a finite number of smaller squares no two the same size.) Among those cubes leading to this squaring there must be a smallest cube. Being smallest it is easy to see that it cannot be located in a corner or on a side of this bottom square. Therefore, it must be in the interior with no edge along an edge of the bottom square. Consider now the top of that cube, in the original filling, that gave rise to this smallest cube. Since this is the cube of smallest side sitting on the bottom of the filled cube the top of this cube is the bottom of a well (i.e., there are walls rising on all sides of this square). In the original filling, the cubes filling this well must all be smaller than the cube whose top is the bottom of the well. By a repetition of the same argument, these cubes induce a squaring on the square at the bottom of the well, etc. Continuing this procedure produces an unending sequence of smaller and smaller cubes and this is not consistent with the original filling having only a finite number of cubes. This completes the argument. (Apparently this was first proved in the paper by Brookes, Smith, Stone, and Tutte, cited below, and was in response to a question raised by Chowla in 1939. An exposition of the result will be found in Honsberger.)

A related, though somewhat different, result is contained in the following quotation from a paper by de Bruijn and Klarner.

> Consider a catalogue S which lists one to infinitely many shapes of rectangular bricks with positive integer dimensions. Using as many bricks of each shape as needed, the bricks listed in S may be used to completely fill certain rectangular boxes. We assume the shapes to be oriented, i.e., we are not allowed to turn bricks around when trying to fill a box. Thus, a new catalogue $\Gamma(S)$ may be formed which lists the (infinitely many) rectangular boxes which may be completely filled with bricks having their shape listed in S. Some of the bricks listed in S may be shapes

> of boxes which may be filled up completely with smaller bricks listed in S; in other words, there may be elements $s \in S$ such that $s \in \Gamma(S) \setminus \{s\}$. The bricks which may be formed with bricks in S smaller than themselves are *composites*. Bricks in S which are not composites are *primes* in S. If $B_\Gamma = B_\Gamma(S)$ is the set of primes in S, then B_Γ is nonempty and every box which can be formed with elements of S can be formed with elements of the subset B_Γ of S; in other words, $\Gamma(B_\Gamma) = \Gamma(S)$ (see Lemma 4). The subject of this note is the remarkable fact that the set of primes $B_\Gamma(S)$ is finite for every set S.

For the proof of this assertion we refer the reader to the original paper, "A Finite Basis Theorem for Packing Boxes with Bricks" [3].

In a discussion of brick-packing problems, Klarner mentions, among other things, that bricks of size $a \times ab \times abc$ may only pack boxes of size $ax \times aby \times abcz$.

As an example he shows that $1\times1\times4$ bricks will not pack a $10\times10\times10$ box. The proof is quite elegant. Coordinatize the cells of the box and then color them with the colors green, blue, red, yellow in accordance with whether the mod 4 remainder of the sum of the coordinates is 0, 1, 2, 3. If the specified bricks could pack the box, then, since in any possible orientation each brick would cover exactly one cell of each color, there would have to be 250 yellow cells, contrary to the fact that there are only 249 yellow cells.

Using an earlier result of de Bruijn—if rectangular bricks will not trivially fill a rectangular box, then they will not fill it at all—the above result is clear since 4 does not divide 10. De Bruijn had discussed the question of whether a $10 \times 10 \times 10$ box could be filled with $1 \times 2 \times 4$ bricks.

At the end of the paper he says "The question of the bricks $1 \times 2 \times 4$ arose from a remark by the proposer's son F. W. de Bruijn who discovered, at the age of 7, that he was unable to fill his $6 \times 6 \times 6$ box by bricks $1 \times 2 \times 4$."

A well-known parlor puzzle is to show that 31 dominos, each capable of covering exactly two squares, may not completely cover a chess board from which two diagonally opposite squares have been removed.

The key to the problem is to observe that each domino covers one square of each color whereas diagonally opposite squares are the same color.

A similar problem, due to Golomb, is to show that 21 3×1 rectangles and 1 1×1 square can cover a chess board (of 1×1 squares) if and only if the 1×1 square is in one of four particular, symmetrically placed, spots.

Mackinnon has given a very elegant combinatorial proof of this which we sketch below. Think of the squares of the chess board as the ordered pairs (a, b), $0 \le a \le 7$, $0 \le b \le 7$, and associate with the pair (a, b) the monomial $x^a y^b$. Further, associate with any set S of the pairs (a, b), i.e., of the squares of the chessboard, the polynomial $\sum_{(a,b)\in S} x^a y^b$. Thus the sets consisting of three consecutive horizontal or three consecutive vertical squares all have associated polynomials of the forms

$$x^i y^j (1 + x + x^2) \qquad \text{or} \qquad x^j y^i (1 + y + y^2).$$

Hence the region covered by the 21 3×1 rectangles has associated polynomial

$$(1 + x + x^2) f(x, y) + (1 + y + y^2) g(x, y),$$

for suitable polynomials $f(x, y)$, $g(x, y)$. If these 21 rectangles covered the entire board with the exception of the square (a, b), then

$$\begin{aligned}(1 + x + x^2)&f(x, y) + (1 + y + y^2)g(x, y) \\ &= (1 + x + x^2 + \cdots + x^7)(1 + y + y^2 + \cdots + y^7) - x^a y^b.\end{aligned}$$

This is a polynomial identity so it must remain true when x and y are replaced by arbitrary complex numbers. If we take $x = y = \omega$, where ω is a non-real cube root of unity, we have

$$(1 + \omega)^2 - \omega^{a+b} = 0 \qquad \text{so} \quad a + b \equiv 1 \pmod 3.$$

If we take $x = \omega$, $y = \omega^2$, we have

$$(1 + \omega)(1 + \omega^2) - \omega^{a+2b} = 1 - \omega^{a+2b} \qquad \text{so} \quad a + 2b \equiv 0 \pmod 3.$$

These two congruences tell us $a \equiv b \equiv 2 \pmod 3$. Therefore the 1×1 square must be one of the four squares $(2,2)$, $(2,5)$, $(5,2)$, $(5,5)$.

With the $(2,2)$ square covered by the 1×1 square we can use 6 horizontal and 15 vertical 3×1 rectangles. The horizontal rectangles are placed with two in the upper left-hand corner, two in the lower right-hand corner, and two just to the left of those in the lower right-hand corner. The rest of the chessboard is then covered with the 15 vertical 3×1 rectangles.

The other three possible positions for the 1×1 square are symmetrically placed with respect to the center of the board so coverings for these are merely rotations of the covering presented.

The "parlor puzzle" mentioned at the start is even more easily handled by this method—but, of course, not as simply as outlined earlier.

See also Integer 23, Filling Boxes and Integer 47, Cutting a Cube into Cubes.

REFERENCES

[1] R. Honsberger, *Ingenuity in Mathematics,* New Math. Library No. 23, Mathematical Association of America, 1970, pp. 56–58.

[2] R. L. Brookes, C. A. B. Smith, A. H. Stone, and W. T. Tutte, The dissection of rectangles into squares, *Duke J.* 77 (1940) 312–340.

[3] N. G. de Bruijn and D. A. Klarner, A finite basis theorem for packing boxes with bricks, *Philips Res. Reports* 30 (1975) 337–343.

[4] D. A. Klarner, Brick-packing puzzles, *J. Rec. Math.* 6 (1973) 112–117.

[5] N. G. de Bruijn, Filling boxes with bricks, *Amer. Math. Monthly* 76 (1969) 37–40.

[6] N. Mackinnon, An algebraic tiling proof, *Math. Gaz.* 73 (1989) 210–211.

3 Sets Congruent to a Proper Part

If one lets θ be incommensurate with π, then the set of points on the circumference of the unit circle whose polar coordinates are $(1, k\theta)$, k a nonnegative integer, form a set E, congruent to a proper part of itself; namely, E is congruent to $E \setminus \{(1,0)\}$ since a rotation counterclockwise by θ takes E into $E \setminus \{(1,0)\}$.

A theorem of Straus tells us that a plane set can have no more than one point such as $(1,0)$ above. (For sets on the line this had been previously proved by Sierpiński.)

On the other hand, in 1955 Mycielski proved that there is an infinite 3-dimensional set congruent to every proper subset obtained from it by the removal of a single point. In fact such a set may be taken on the surface of a sphere.

See also Integer 5, The Banach–Tarski Paradox.

REFERENCES

[1] E. G. Straus, On a problem of Sierpiński on the congruence of sets, *Fund. Math.* 44 (1957) 75–81.

[2] J. Mycielski, About sets with strange isometric properties, *Fund. Math.* 42 (1955) 1–10.

4 Random Walks

In *An Introduction to Probability Theory and its Applications,* W. Feller gives the following statement of a theorem due to Pólya.

Theorem (Pólya). *In the symmetric random walks in one and two dimensions there is probability one that the particle will sooner or later (and therefore infinitely often) return to its initial position. In three dimensions, however, this probability is* < 1. (p. 360)

The probability of return in three dimensions is approximately 0.35.

REFERENCES

W. Feller, *An Introduction to Probability Theory and its Applications,* (3rd ed.), Wiley, New York, 1968.

Integer 3

1 Sums

For p and q positive integers and c constant, Edmonds has shown that

$$\sum_{n=1}^{N} cn^p = \left(\sum_{n=1}^{N} n^q\right)^2 \qquad \text{for all } N \geq 1$$

is impossible except for $c = 1, p = 3, q = 1$.

In the exceptional case, we have

$$1^3 + 2^3 + \cdots + n^3 = (1 + 2 + \cdots + n)^2.$$

This special identity is also given by taking $m = p^n$ in the identity (going back to Liouville—$\tau(n)$ is the number of divisors of n)

$$\left(\sum_{d|m} \tau(d)\right)^2 = \sum_{d|m} \tau^3(d).$$

Mąkowski showed these to be the only integers k, ℓ, p, $\ell > 1$, such that

$$\left(\sum_{d|m} \tau^k(d)\right)^\ell = \sum_{d|m} \tau^p(d).$$

In the review of Edmonds paper (*Math. Rev.* 20 #3098) Straus observed that even $1 + 2^p = (1 + 2^q)^r$, $r > 1$, has no solutions other than $p = 3$, $q = 1$, $r = 2$.

Utz has asked for solutions of

$$x_1^3 + x_2^3 + \cdots + x_n^3 = (x_1 + x_2 + \cdots + x_n)^2$$

in positive integers x_j and has shown that for each fixed n this equation has only finitely many such solutions. However, the number of solutions becomes unbounded as n tends to infinity.

See also Integer 4, A Few Diophantine Equations and Integer 8, Square Identities.

REFERENCES

[1] S. M. Edmonds, Sums of powers of the natural numbers, *Math. Gaz.* 41 (1957) 187–88.

[2] A. Mąkowski, Remarques sur les fonctions $\theta(n)$, $\phi(n)$ et $\sigma(n)$, *Mathesis* 69 (1960) 302–303.

[3] W. R. Utz, The Diophantine equation $(x_1 + \cdots + x_n)^2 = x_1^3 + \cdots + x_n^3$, *Fib. Quart.* 15 (1977) 14, 16.

2 Beatty's Theorem

Let $S(\alpha) = \{[\alpha], [2\alpha], [3\alpha], \ldots\}$, where $[x]$ is the largest integer not exceeding x.

Theorem 2.1 (Beatty). *If α is positive and irrational and if $\frac{1}{\alpha} + \frac{1}{\beta} = 1$, then every positive integer is in exactly one of the sequences $S(\alpha)$, $S(\beta)$.*

In fact, from $\frac{1}{\alpha} + \frac{1}{\beta} = 1$, we find

$$n-2 = \left(\frac{n}{\alpha}-1\right) + \left(\frac{n}{\beta}-1\right) < \left[\frac{n}{\alpha}\right] + \left[\frac{n}{\beta}\right] < \frac{n}{\alpha} + \frac{n}{\beta} = n$$

and, therefore, $[\frac{n}{\alpha}] + [\frac{n}{\beta}] = n-1$. Since the left-hand terms are the respective numbers of multiples of α and of β not exceeding n, and since this is true for all n Beatty's theorem follows.

The following is a theorem of Uspensky (see Uspensky [9] or Graham [5]).

Theorem 2.2 (Uspensky). *There do not exist 3 or more numbers $\alpha_1, \alpha_2, \ldots, \alpha_n$ such that $S(\alpha_1), \ldots, S(\alpha_n)$ are nonempty disjoint sequences which, taken together, contain each positive integer precisely once.*

On the other hand, Beatty's theorem may be used to give other types of exact splittings of the set of positive integers into three or more sets as follows. For $\tau = \frac{1+\sqrt{5}}{2}$, we have $\tau^2 = \tau + 1$ so $1 = \frac{1}{\tau} + \frac{1}{\tau^2}$ and

$$\{[n\tau] \mid n = 1,2,3,\ldots\}, \qquad \{[n\tau^2] \mid n = 1,2,3,\ldots\}$$

is an exact splitting of the positive integers. Using this we see that

$$\{[[n\tau]\,\tau] \mid n = 1,2,3,\ldots\}, \qquad \{[[n\tau^2]\,\tau] \mid n = 1,2,3,\ldots\}$$

exactly exhausts $\{[n\tau] | n = 1,2,\ldots\}$. Hence

$$\{[[n\tau]\,\tau]\}, \qquad \{[[n\tau^2]\,\tau]\}, \qquad \{[n\tau^2]\}$$

exactly exhaust the positive integers. This splitting was first given by Skolem.

Continuing in a similar way, if we define

$$A_0 = S(\tau), \qquad A_{m+1} = \{[n\tau^2] \mid n \in A_m\},$$

then these sets $A_0, A_1, A_2, \ldots$ exactly exhaust the positive integers. (These remarks may be generalized to $\frac{1}{\alpha} + \frac{1}{\beta} = 1$.) As an amusing aside one may note that $[n\tau] = n + g(n-1)$, where g is defined by

$g(0) = 0$, $g(n) = n - g(g(n-1))$. This was proved by Granville and Rasson in their article "A Strange Recursive Relation."

Golomb (in 1976) gave the following exact splitting of the set of positive integers. Let $\{q_n\}$ be a sequence of integers such that $1 \le q_0 < q_1 < q_2, \ldots$ and put $\pi(n)$ for the number of terms in the sequence not exceeding n. Noting that, for each j,

- $q_j + \pi(q_j), \ldots, q_j + 1 + \pi(q_j + 1), \ldots, (q_{j+1} - 1) + \pi(q_{j+1} - 1)$ are consecutive integers;
- $q_{j+1} + \pi(q_{j+1}) = (q_{j+1} - 1) + \pi(q_{j+1} - 1) + 2;$
- $(q_{j+1} - 1) + \pi(q_{j+1} - 1) + 1 = q_{j+1} + j;$

we see that

$$\{n + \pi(n) | n = 1, 2, \ldots\}, \qquad \{q_n + n - 1 | n = 1, 2, \ldots\}$$

is an exact splitting of the positive integers.

Golomb deduced Beatty's theorem from this result.

This splitting was essentially given by Lambek and Moser in 1954 and they note in a postscript that similar ideas go back at least to Brun and Lehmer who used them in 1931 and 1932.

Lambek and Moser give a further result which has the following amusing special cases.

The nth nonprime is the limit of the sequence

$$n, n + \pi(n), n + \pi(n + \pi(n)), \ldots.$$

The nth positive integer which is not a perfect kth power ($k \ge 2$) is

$$n + \left[\left(n + \left[n^{1/k}\right]\right)^{1/k}\right].$$

The nth positive integer not of the form $[e^m]$, with $m \ge 1$, is

$$n + [\log(n + 1 + [\log(n + 1)])].$$

This sort of problem has recently been revisited by Dos Reis and Silberger.

An unrelated result involving the quantities $[n\tau]$, where τ is as above, is the following remarkable expansion due to Davison.

$$\frac{1}{2^{[\tau]}}+\frac{1}{2^{[2\tau]}}+\frac{1}{2^{[3\tau]}}+\cdots=\cfrac{1}{2^0+\cfrac{1}{2^1+\cfrac{1}{2^1+\cfrac{1}{2^2+\cfrac{1}{2^3+\cfrac{1}{2^5+\cfrac{1}{2^8+\cfrac{1}{2^{13}+\cdots}}}}}}}}.$$

REFERENCES

[1] S. Beatty, Problem 3173, *Amer. Math. Monthly* 34 (1927) 159.

[2] R. L. Graham, On a theorem of Uspensky, *Amer. Math. Monthly* 70 (1963) 407–409.

[3] J. V. Uspensky, On a problem arising out of the theory of a certain game, *Amer. Math. Monthly* 34 (1927) 516–21.

[4] J. B. Roberts, *Elementary Number Theory,* MIT Press, Boston, 1977, pp. 39–40.

[5] V. Granville and J. P. Rasson, A strange recursive relation, *J. of Number Theory* 30 (1988) 238–241.

[6] S. Golomb, The "sales tax" theorem, *Math. Mag.* 49 (1976) 187–89.

[7] J. Lambek and L. Moser, Inverse and complementary sequences of natural numbers, *Amer. Math. Monthly* 61 (1954) 454–58.

[8] A. J. Dos Reis and D. M. Silberger, Generating non-powers by formula, *Math. Mag.* 63 (1990) 53–55.

[9] J. L. Davison, A series and its associated continued fraction, *Proc. Amer. Math. Soc.* 63 (1977) 29–32.

[10] W. W. Adams, J. L. Davison, A remarkable class of continued fractions, *Proc. Amer. Math. Soc.* 65 (1977) 194–98.

3 Iterates

Let $a_1a_2\cdots a_p$ be the decimal representation of n and put $f(n) = a_1^3 + a_2^3 + \cdots + a_p^3$. Then there is an m_0 such that for $m \geq m_0$ the mth iterate of f applied to n is equal to 153 (i.e., $f^{(m)}(n) = 153$) if and only if 3 is an integral divisor of n.

A well-known observation of Steinhaus is that if one defines the function $g(n)$ to be the sum of the squares of the base 10 digits of n, rather than the sum of the third powers of the base 10 digits as above, then, if one iteratively applies this function, one always either comes to 1 or to the cycle $(145, 42, 20, 4, 16, 37, 58, 89)$.

Thus iteration of g applied to an integer n will always lead to 1 or to 4. Crittenden and Harris have shown that there can be arbitrarily long gaps between consecutive n leading to the same one of these two values.

If one defines a function, such as g above, where one raises the digits to the kth power, where $k \geq 2$, then iteration will always lead to a finite number of cycles. For proofs of this, see either Kiss, Hasse and Pritchett, or Burkard.

The observation in the first sentence of this subsection is due to Matsuoka and tells us that when $k = 3$ all multiples of 3 end in the cycle (153) of length 1.

In the Hasse–Prichett paper, they prove:

> Let $\phi(\alpha)$ be a polynomial with positive integral values for $\alpha = 0, 1, \ldots, g-1$ and let a be a natural number with $n \geq 1$ g-adic digits,
>
> $$a = \alpha_0 + \alpha_1 g + \cdots + \alpha_{n-1} g^{n-1},$$
>
> $0 \leq \alpha_\nu \leq g-1$, $\nu = 0, 1, \ldots, n-1$; $\alpha_{n-1} \geq 1$, and define
>
> $$a' = \phi(\alpha_0) + \phi(\alpha_1) + \cdots + \phi(\alpha_{n-1}).$$

> If one repeats this procedure indefinitely to form a sequence $a, a', a'', \ldots$, then this sequence will eventually become periodic. [4]

The periodic portion of the sequence of iterates of a will be called a *g-adic cycle* of ϕ.

In the case where $\phi(x) = x^2$ it is shown that: the set of cycles determined by all a less than or equal to g^2 is a complete set of cycles for the g-adic integers.

Hasse and Prichett asked when there will be a single cycle of length greater than one. Using the above result and a computer, they formulated the conjecture that the only possible g, permitting exactly one such cycle, are given by $g = 6, 10, 16, 20$.

The mathematical literature is filled with similar questions concerning the iteration of various functions. We shall confine ourselves to two further instances here.

An old question of Catalan asks if the sequence of iterates of the function

$$f(n) = \sigma(n) - n, \qquad f(1) = 1.$$

is periodic. The answer is not known. Note that the non-prime values of n (> 1) for which the sequence is purely periodic and with period 1 are precisely the perfect integers. Assuming that every even number greater than 8 is a sum of two distinct primes Sierpiński has shown that there are infinitely many integers whose sequence of iterates ends with the cycle (6). The numbers 12496 and 14316 lead to cycles of lengths 4 and 28. The numbers in the amicable pair 220, 284 each go to the other. It is known that there are integers n for which the iteration is as long as one might wish. Guy and Selfridge have conjectured that far from being periodic the iterates go off to infinity infinitely often. (See also Integer 5, A Conjecture Based on Goldbach's Conjecture.)

Not infrequently one is asked to "find the next term of the sequence starting . . . ." Often the underlying rule depends upon finding a function whose iterates generate the sequence in question. An example of this is given by Mordell in a review of a book by Sierpiński. In the review he writes:

> I recall a problem given to me perhaps about fifteen years ago

> ... to find the next term of the sequence
>
> $$4, 8, 21, 52, 65, 96.$$
>
> It may be added that the solution involves only the simplest arithmetic and that the answer is striking. When I gave this to Prof. Sierpiński he solved the problem and then investigated the periodicity of similar sequences. It may be noted that the 61st term of the sequence is the same as the asked for seventh term. [9]

In this case, though Mordell leaves the reader in the air about it, the next term is 1 and the next few terms beyond that are

$$5, 9, 31, 53, 75, 97, 101, 501.$$

The rule which makes all this clear is to start with 4 and then at each step add 4 and reverse the digits.

Again we refer to Burkard for an entrée to the literature.

REFERENCES

[1] Y. Matsuoka, On a problem in number theory, Sci. Rep. Kagoshima Univ. No. 14 (1965) 7–11.

[2] R. B. Crittenden and J. K. Harris, A variation on a problem in number theory of H. Steinhaus, *Elem. Math.* 21 (1966) 60.

[3] P. Kiss, A generalization of a problem in number theory, Math. Sem. Notes Kobe Univ. 5 (1977) 313–317.

[4] H. Hasse and G. Prichett, A conjecture on digital cycles, *J. für Reine und Angew. Math.* 298 (1978) 8–15.

[5] R. E. Burkard, Iteriertezahlen theoretische Funktionen, *Math.-Phys. Semesterberichte* 19 (1971) 235–253.

[6] W. Sierpiński, Sur les itérations des certaines fonctions numériques, *Rend. Circ. Mat. Palermo* 13 (1964) 257–262.

[7] R. K. Guy and J. L. Selfridge, What drives an aliquot sequence?, *Math. Comp.* 29 (1975) 101–107. (Corrigendum, same journal 34 (1980) 319–321.)

[8] R. K. Guy, Aliquot sequences in *Number Theory and Algebra* (ed. H. Zassenhaus), Academic Press, New York, 1977, pp. 111–118.

[9] L. J. Mordell, Review of *A Selection of Problems in the Theory of Numbers* by W. Sierpiński, *J. London Math. Soc.* 40 (1965) 571–573.

4 $\sigma(p^4)$

The only prime p such that $\sigma(p^4)$ is a square is 3. That is, $1 + p + p^2 + p^3 + p^4$ is a square, when p is a prime, precisely when $p = 3$. In that case we have

$$1 + 3 + 3^2 + 3^3 + 3^4 = 11^2.$$

See also Integer 7, A Square.

REFERENCES

V. Thébault, Curiosités arithmétiques, *Mathesis* 62 (1953) 120–129.

5 Chaos—An Ex-"Property of 3"

Let F be a continuous map of an interval J into itself. Suppose there is a point a in J for which the points $b = F(a)$, $c = F^2(a)$, $d = F^3(a)$ (i.e., $a \to b \to c \to d$), satisfy $d \leq a < b < c$ or $d \geq a > b > c$. Then

1. for every $n = 1, 2, \ldots$ there exists a point in J with period n;
2. there is an uncountable subset S of J such that:
   a. for all p,q in S with $p \neq q$,
      i. $\limsup_n |F^n(p) - F^n(q)| > 0$ and

ii. $\liminf_n |F^n(p) - F^n(q)| = 0$;

b. for every p in S and periodic point q in J

$$\limsup_n |F^n(p) - F^n(q)| > 0.$$

If there is a point of period 3, i.e., a point a such that $a \to b \to c \to a$, then the hypotheses of the theorem are satisfied.

It was later shown that if a point in J, J compact, has period divisible by any of 3, 5, or 7, then F is chaotic.

Later yet, under the same conditions, it was shown that if a point has a period other than a power of 2, then F is chaotic. Further, if p is a power of 2, then there is an F, with a point of period p, which is not chaotic.

REFERENCES

[1] T. Li and J. A. Yorke, Period three implies chaos, *Amer. Math. Monthly* 82 (1975) 985–992.

[2] M. B. Nathanson, Permutations, periodicities, and chaos, *J. Comb. Thy. A* 22 (1977) 61–8.

[3] F. J. Fuglister, A note on chaos, *J. Comb. Thy. A* 27 (1979) 186–88.

6 Repeating Blocks

The longest string of 2 symbols, say 0 and 1, that does not have adjacent identical blocks of digits (of any length) is exemplified by 010. This is easily seen since taking (without loss of generality) the first digit to be zero we see that the next must then be 1 and the third must be 0 again. Now, however, placing a 0 in the 4th position produces a repeating block of length 1, namely 0, while placing a 1 in the 4th position produces a repeating block of length 2, namely 01.

If we ask for the longest string of 3 symbols, say 0, 1, and 2, that does not have identical adjacent blocks of digits (of any length), it is somewhat surprising to learn that finite strings of all lengths exist meeting this condition. In fact one can even construct an infinite string meeting the condition.

First we construct an infinite string of 0's and 1's. We start with a 0 then we put the "opposite" of 0, the 1, then the "opposites" of these two, namely 10, etc. (We put a vertical bar each time we write the "opposites" of the preceding terms.)

0|1|10|1001|10010110|1001011001101001|10010110011010010110100110010...

We write $a_0, a_1, \ldots$ for the terms of the resulting sequence and note that those terms just following one of the vertical bars are of the form a_{2^n}. Thus the typical term in the stretch between two consecutive vertical bars is a_{2^n+j}, where $0 \leq j < 2^n$. By the construction this implies $a_j \neq a_{2^n+j}$. (For example, since $a_0 = 0$ we see that $a_{2^n} = 1$ for all n.) Applying the rule t times we see, when $n_1 < n_2 < \cdots < n_t$, that $a_{2^{n_t}+\cdots+2^{n_1}}$ is 0 when t is even and is 1 otherwise. I.e., a_n is 0 when there are an even number of 1's in the binary expansion of n and is 1 otherwise.

It is immediate that $a_{2n} = a_n$ and $a_{2n} \neq a_{2n+1}$. From this we see, since exactly one of $j, j+1$ is even, that no three consecutive terms in the a_n sequence, say a_j, a_{j+1}, a_{j+2}, can all be equal to 1.

Thus there will never be more than two 1's between consecutive 0's in this sequence. We now construct a new sequence by letting its jth term be the number of 1's between the jth and the $j+1$st zero terms of the a_n sequence. Using the above segment of the a_n sequence we obtain:

2 1 0 2 0 1 2 1 0 1 2 0 2 1 0 2 0 1 2 0 2 1 0 1 2 1 0 2 0 1....

This is an infinite sequence with no two identical adjacent blocks. To see this we assume the sequence contains a string of the form EE, where E is a block of digits. This implies (by the way the sequence was obtained from the earlier sequence) that the original sequence of 0's and 1's contains a string of the form FFf, where f is the first digit of the block of digits F. We will show this is not possible—thereby proving the desired result.

Assuming such a string exists we let FFf, where

$$F = a_{i+1}a_{i+2}\cdots a_{i+k},$$

be the leftmost instance. If k were odd, then, since $a_{i+j} = a_{i+k+j}$ for $1 \le j \le k$ and since exactly one of $i+j, i+j+k$ is even, the elements of F alternate in value between 0 and 1. Thus, for odd k, F has an even number of alternations and, therefore,

$$a_{i+1} = a_{i+k} = a_{i+k+1} = a_{i+2k} = a_{i+2k+1}.$$

But when i is odd the 2nd equality is false and when i is even the 4th equality is false. Thus k is even.

With k even and i even we note

$$a_i FFf = F'F'f'a_{i+1+2k},$$

where $F' = a_i a_{i+1} \cdots a_{i+k-1}$, so that FFf was not the earliest of its type. Thus i must be odd.

Suppose $i = 2s - 1$ and $k = 2t$. Then, using $a_m = a_{2m}$ for all m, we see that $F''F''f''$, where

$$F'' = a_s a_{s+1} \cdots a_{s-1+t},$$

precedes FFf, again contradicting the assumption that FFf was the earliest. This completes our proof.

The first such sequence was constructed by Thue in 1906 and it has been continuously rediscovered ever since.

In the mid-forties Morse and Hedlund rediscovered the sequence while investigating a proposed scheme for an "ending rule" in chess.

Dekking [8] showed the (perhaps remarkable) fact that if one places a decimal point at the beginning of the sequence of 0's and 1's displayed above, then the resulting (base 2) real number is transcendental.

Entringer and Jackson proved the following about *binary* sequences.

1. There exists an infinite binary sequence with no identical adjacent blocks of length greater than or equal to 3.

2. Every binary sequence of length greater than 18 has identical adjacent blocks of length greater than or equal to 2.

3. Every infinite binary sequence has arbitrarily long adjacent blocks that are permutations of each other. (If binary is replaced by k-ary, $3 \leq k \leq 5$, see Brown.)

In yet a different paper, Dekking [11] proved the following.

1. There are infinite binary sequences having no triple repetitions and no repetitions of length 4 or greater.
2. Binary sequences with no triple repetitions and no repetitions of length 3 or greater are finite.
3. Infinite binary sequences with no identical overlapping blocks have arbitrarily long identical adjacent blocks.

The sequence 0110100110010110100101 10... has an interesting arithmetic property first observed by Prouhet in 1851. Under the first 2^n terms of this sequence write the first 2^n consecutive integers (or, if you like, some sequence of 2^n consecutive terms of any arithmetic progression). Then this sequence of numbers is split into two classes—one of them consisting of those numbers below a 0 and the other consisting of those numbers below a 1. For example, for $n = 4$, we would have the following pair of rows.

$$\begin{array}{cccccccccccccccc} 0 & 1 & 1 & 0 & 1 & 0 & 0 & 1 & 1 & 0 & 0 & 1 & 0 & 1 & 1 & 0 \\ 1 & 2 & 3 & 4 & 5 & 6 & 7 & 8 & 9 & 10 & 11 & 12 & 13 & 14 & 15 & 16 \end{array}$$

This gives the two sets of integers A, B, where

$$A = \{1, 4, 6, 7, 10, 11, 13, 16\}, \qquad B = \{2, 3, 5, 8, 9, 12, 14, 15\}.$$

Doing this for a general n always results in two sets such that the jth powers of the numbers in the first set add to the same sum as the jth powers of the numbers in the second set, for all j, $0 \leq j \leq n - 1$. (See also Integer 13, Three Thirteens.)

In our example, in the case $n = 4$, we have the sum of the jth powers of the numbers in A equal to the sum of the jth powers of the numbers in B, for $j = 0, 1, 2, 3$.

Similar splittings exist for dividing consecutive integers into 3 or 4 or 5 or ... classes with equal power sums. We illustrate this for the case of 3. Construct the following finite strings of 0's, 1's, and 2's carefully noting the pattern of passage from one "line" to the next. (The last two lines constitute the 4th "line".)

012
012 120 201
012120201 120201012 201012120
012120201120201012201012120 120201012201012120012120201
201012120012120201120201012
...

Writing under the successive rows the consecutive integers 1 to 3 for the first row, 1 to 9 for the second row, 1 to 27 for the third row, 1 to 81 for the fourth row, etc. we see that, in the respective cases, these integers are split into the following sets of three classes each.

1; 2; 3
1,6,8; 2,4,9; 3,5,7
1,6,8,12,14,16,20,22,27; 2,4,9,10,15,17,21,23,25; 3,5,7,11,13,18,19,24,26
...

Considering the nth row of this array (starting with $n = 1$ for the top row) one finds that the sums of the tth powers of the integers in each of the three classes of that row are the same for all values of $t < n$.

Thus for $t = 0, 1, 2$, respectively, the sums of the tth powers of 1, 6, 8, 12, 14, 16, 20, 22, 27 are 9, 126, 2310 and the same is true for the sums of the tth powers of 2, 4, 9, 10, 15, 17, 21, 23, 25 and for the tth powers of 3, 5, 7, 11, 13, 18, 19, 24, 26. There is a large literature on the subjects of the above observations.

Lambek and Moser used (essentially!) the sequence

0110100110010110...

to split the entire set of nonnegative integers into the two disjoint and mutually exhaustive classes

$$\{0, 3, 6, 9, 10, 12, 15, \ldots\} \quad \text{and} \quad \{1, 2, 4, 8, 7, 11, 13, 14, \ldots\}.$$

They then showed that the set of numbers obtained by adding any two distinct elements of the first class is exactly the same as the set of numbers obtained by adding any two distinct members of the second class. Furthermore, if a number is the sum of distinct members of the first class in more than one way, it is obtainable exactly the same number of ways by adding elements of the second class. Finally, they showed that no other splitting of the set of all nonnegative integers into two classes has these properties.

Stopping the sequence after 2^k terms yields a splitting with the same properties but stopping the sequence at any other point fails to yield such a splitting. They asked if it is possible to split all nonnegative integers into three classes having the same property and added that they do not know the answer.

Lambek and Moser also gave the splitting

$$C = \{1, 6, 8, 10, 12, 14, 15, 18, 20, 21, 22, 26, 27, 28, \ldots\},$$

$$D = \{2, 3, 4, 5, 7, 9, 11, 13, 16, 17, 19, 23, 24, 25, 29, 30, \ldots\}$$

which has the above mentioned property for multiplication rather than for addition. The rule for determining whether an integer n goes into C or D is as follows. If the canonical factorization of n into prime powers has the exponents $r_1, \ldots, r_k$, then n is in C if the total number of 1's in the binary representations of the r_j is even and is in D otherwise.

No such splitting of the first n nonnegative integers for multiplication is possible.

Since that time there has been some interest in knowing when the set of subset sums for a finite set of integers will determine the set of integers. We refer the reader to the paper by Ewell, "On the Determination of Sets by Sets of Sums of Fixed Order."

We confine our remaining remarks in this subsection to a result proved by Shelton.

From the sequence 2102012101202102..., by placing a decimal point before the leading 2, we have the base 3 expansion of a real number between 0 and 1. Let S be the totality of all real numbers obtainable from such sequences of 0's, 1's, and 2's (with no identical adjacent blocks). Then the set S is *perfect* (i.e., it is closed, bounded, and has no isolated points).

A reader who finds these sequences interesting might be stimulated by the series of articles by Dekking, Mendés France, and van der Poorten which go under the title"Folds!."

REFERENCES

[1] A. Thue, Über unendliche Zeichenreihen, Norske Vid. Selsk. Skr. I Mat.-Nat. Kl. 7, 1906, pp. 1–22. (*Selected Mathematical Papers* (eds. T. Nagell, et al.) Universitetsforlaget, Oslo-Bergen-Tromsø, 1977, pp. 139–158.)

[2] ——, Über die gegenseitigen Lage gleicher Teile gewisser Zeichenreihen, Ibid., 1912, pp. 1–67. (*Selected Mathematical Papers* (eds. T. Nagell, et al.) Universitetsforlaget, Oslo-Bergen-Tromsø, 1977, pp. 413–478.)

[3] ——, Über Veränderungen von Zeichenreihen nachgegebenen Regeln, Ibid., 1914, pp. 493–524. (*Selected Mathematical Papers* (eds. T. Nagell, et al.) Universitetsforlaget, Oslo-Bergen-Tromsø 1977, pp. 493–524).

[4] G. A. Hedlund, Remarks on the work of Axel Thue on sequences, *Nord. Mat. Tids.* 15 (1967) 148–150.

[5] C. H. Braunholtz, An infinite sequence of three symbols with no adjacent repeats, *Amer. Math. Monthly* 10 (1963) 675–76.

[6] P. A. B. Pleasants, Non-repetitive sequences, *Proc. Camb. Phil. Soc.* 68 (1970) 267–274.

[7] M. Morse and G. A. Hedlund, Unending chess, symbolic dynamics and a problem in semi-groups, *Duke J. Math.* 11 (1944) 1–7.

[8] F. M. Dekking, Transcendance du nombre de Thue–Morse, *C. R. Acad. Sci. Paris Sér. A-B* 285 (1977) 157–160.

[9] T. C. Brown, Is there a sequence on four symbols in which no two adjacent segments are permutations of one another?, *Amer. Math. Monthly* 78 (1971) 886–88.

[10] R. C. Entringer and D. E. Jackson, On nonrepetitive sequences, *J. Comb. Thy. series A* 16 (1974) 159–164.

[11] F. M. Dekking, On repetition of blocks in binary sequences, *J. Comb. Thy. series A* 20 (1976) 292–99.

[12] M. E. Prouhet, Mémoire sur quelque relations entre les puissances des nombres, *C. R. Acad. Sci.* 33(1851) 225.

[13] D. H. Lehmer, The Tarry–Escott problem, *Scripta Math.* 13 (1947) 37–41.

[14] J. B. Roberts, A curious sequence of signs, *Amer. Math. Monthly* 64 (1957) 317–322.

[15] ——, Splitting consecutive integers into classes with equal power sums, *Amer. Math. Monthly* 71 (1964) 25–37.

[16] ——, Relations between the digits of numbers and equal sums of like powers, *Can. J. Math.* 16 (1964) 626–636.

[17] J. Lambek and L. Moser, On some two way classifications of integers, *Can. Bull. Math.* 2 (1959) 85–89.

[18] J. A. Ewell, On the determination of sets by sets of sums of fixed order, *Can. J. Math.* 20 (1968) 596–611.

[19] R. Shelton, Aperiodic words on three symbols, *J. für Reine und Angew. Math.* 321 (1981) 195–209.

[20] ——, Aperiodic words on three symbols, II, *J. für Reine und Angew. Math.* 327 (1981) 1–11.

[21] R. Shelton and R. P. Soni, Aperiodic words on three symbols, III, *J. für Reine und Angew. Math.* 330 (1982) 44–52.

[22] H. M. Morse, Recurrent geodesics on a surface of regular curvature, *Trans. Amer. Math. Soc.* 22 (1921) 84–100.

[23] M. Dekking, M. Mendés France, and A. J. van der Poorten, Folds!, *The Mathematical Intelligencer* 4 (1982) 130–38, 173–181, 190–95.

7 Sectioning a Circle

Let C be a circle of circumference 1 and center at the origin and let A be an arbitrary point on C. Let ξ be an irrational number and, for each positive integer j, let the point P_j be that point on C at a counterclockwise distance from A of $j\xi$. Further let $A_1, \ldots, A_n$ be the consecutive points $P_1, \ldots, P_n$ in clockwise order, starting with A. (Note that no matter how many times the points P_j, as j goes from 1 to n, may travel the entire circumference of the circle the points A_j, as j goes from 1 to n, never traverse the entire circumference more than once.)

The following was conjectured by Steinhaus and first proved by Świerczkowski in 1958.

Conjecture (Steinhaus). *For no n are there more than 3 different size gaps between consecutive elements of the finite sequence $A_1, \ldots, A_n$.*

This is true despite the fact that the infinite sequence of points $P_1, P_2, \ldots$ is dense on the circumference C.

Chung and Graham generalized the assertion in the following way.

Let d, $n_1, \ldots, n_d$ be positive integers and θ, $\alpha_1, \ldots, \alpha_d$ be arbitrary real numbers. Consider the $N = n_1 + \cdots + n_d$ numbers (where $[x]$ is the largest integer not exceeding x)]

$$k\theta + \alpha_i - [k\theta + \alpha_i], \qquad 1 \leq k \leq n_i,\ 1 \leq i \leq d,$$

and let them be denoted by $x_1, \ldots, x_N$, where $x_1 < \cdots < x_N$. Then there are at most $3d$ distinct differences $x_{i+1} - x_i$.

The Świerczkowski result is the case $d = 1$.

In 1979 Liang gave a simple proof of Graham's result. In his review of the paper Graham says "In this note the author provides a very clean and pleasing proof of the general result."

More recently van Ravenstein has again considered the problem.

REFERENCES

[1] S. Świerczkowski, On successive settings of an arc on the circumference of a circle, *Fund. Math.* 46 (1958–59) 187–89.

[2] F. R. K. Chung and R. L. Graham, On the set of distances determined by the union of arithmetic progressions, *Ars Comb.* 1 (1976) 57–76.

[3] F. M. Liang, A short proof of the 3d distance theorem, *Disc. Math.* 28 (1979) 325–26.

[4] T. van Ravenstein, The three-gap theorem (Steinhaus conjecture), *J. Aust. Math. Soc.* 45 (1988) 360–370.

8 k-transposable Integers

Given a positive integer k another positive integer x is said to be k-*transposable* if, when its leftmost digit is moved to the unit's place, the resulting integer is kx.

The integer 142857 is 3-transposable since

$$428571 = 3 \cdot 142857.$$

Kahan proved: for $k = 1$ all digits of a k-transposable integer must be the same and for $k > 1$ there are no such integers unless $k = 3$ and then they are all in one of the following two sequences.

$$142857,\ 142857142857,\ 142857142857142857, \ldots$$
$$285714,\ 285714285714,\ 285714285714285714, \ldots.$$

In a later paper Kahan considered the problem of moving a digit from the right to the left.

Anne Ludington generalized the question to include bases g other than 10. She showed that there is a base g n-digit k-transposable integer if and only if there is an integer d greater than and relatively prime to k such that $d | g - k$ and $k^n \equiv 1 \pmod{d}$. She also showed that for $g = 5$ or $g \geq 7$ there do exist integers which are k-transposable for suitable k but when $k > 2$ no such k-transposable integers exist when $g = 2, 3, 4, 6$.

Wlodarski asked for the smallest base 10 integer with unit's digit 6 such that if that digit were shifted to the far left, the resulting number would be 6 times the original and gave the solution

$$1016949152542372881355932203389830508474576271186440677966;$$

an integer having 58 digits.

The more general question arising when one transfers digits from the right to the left is discussed in a paper by Bernstein, "Multiplicative Twins and Primitive Roots."

Sutcliffe asserts that the only integers smaller than 10000 which become multiples of themselves when their digits are reversed are: 9801 =

$9 \cdot 1089$, and $8712 = 4 \cdot 2178$. (For further details see Integer 4, Factorials.)

Many other interesting digit patterns may be found in any of the numerous recreational books in mathematics, for example, Coxeter and Rouse Ball or Madachy.

REFERENCES

[1] S. Kahan, k-transposable integers, *Math. Mag.* 49 (1976) 27–8.

[2] ——, k-reverse transposable integers, *J. Rec. Math.* 9 (1976/77) 15–20.

[3] Anne Ludington, Transposable integers in arbitrary bases, *Fib. Quarterly* 25 (1987) 263–67.

[4] J. Wlodarski, A number problem, *Fib. Quarterly* 9 (1971) 195, 198.

[5] L. Bernstein, Multiplicative twins and primitive roots, *Math. Zeit.* 105 (1968) 49–58.

[6] A. Sutcliffe, Integers that are multiplied when their digits are reversed, *Math. Mag.* 39 (1966) 282–87.

[7] H. S. M. Coxeter and W. W. Rouse Ball, *Mathematical Recreations and Essays* 12 ed., U. of Toronto Press, Toronto, 1974.

[8] J. S. Madachy, *Mathematics on Vacation*, Scribners, New York, 1966.

9 Kuratowski's Theorem

A necessary and sufficient condition for a graph to be planar is that neither of the two graphs consisting of:

1. 5 points each joined to every other by an edge;
2. 6 points some 3 of which are each joined to all of the remaining 3;

is a subgraph of the graph.

Integers 3, 4

I Lattice Points in Tetrahedra

When $a_1, \ldots, a_m$ are positive integers we put $N(a_1, \ldots, a_m)$ for the number of lattice points $(x_1, \ldots, x_m)$ in the m-dimensional tetrahedron described by $0 \le x_i \le a_i$ $(1 \le i \le m)$, $0 < \sum_{i=1}^{m} \frac{x_i}{a_i} < 1$.

In 1963 Rademacher conjectured that when the a_j were relatively prime in pairs then

$$N(a_1, \ldots, a_m) \equiv \frac{\prod_{i=1}^{m}(a_i + 1)}{2^{m-1}} \pmod{2}.$$

For $m = 1, 2$ this is easy to see and for the case $m = 3$ Rademacher and Artuhov (independently) proved it to be the case.

In 1979 Rosen showed that there are infinitely many counterexamples to the conjecture when $m = 4$. Consequently, the conjecture is false for $m \ge 4$.

REFERENCES

K. H. Rosen, Lattice points in four-dimensional tetrahedra and a conjecture of Rademacher, *J. für die Reine und Angew. Math.* 307/308 (1979) 264–275.

2 A Problem of Mnich

In 1936 Sierpiński propounded the following problem, attributed to W. Mnich.

> Do there exist three rational numbers whose sum and product are each equal to 1?

The problem is equivalent to the existence of a rational number r such that $x^3 - x^2 + rx - 1 = 0$ has only rational solutions.

In 1960 Cassels traced the roots of the problem to the work of Sylvester and proved that the answer is no. Later he and Sansone considerably simplified the proof.

In 1979 Bohigian showed that for $k > 3$ there are infinitely many sets of rational numbers x_j such that

$$x_1 + \cdots + x_k = x_1 \cdots x_k = 1.$$

In the case of $k = 4$, Schinzel gave

$$\frac{n^2}{n^2-1},\ \frac{1}{1-n^2},\ \frac{n^2-1}{n},\ \frac{1-n^2}{n}.$$

In the cases $k = 5, 6, 7$, Bohigian gave, respectively,

$$n,\ \frac{-1}{n},\ -n,\ \frac{1}{n},\ 1$$

$$\frac{1}{n^2(n+1)},\ \frac{-1}{n^2(n+1)},\ (n+1)^2,\ -n^2,\ -n,\ -n$$

$$(n-1)^2,\ n-\frac{1}{2},\ n-\frac{1}{2},\ 1,\ -n^2,\ \frac{1}{n(n-1)(n-\frac{1}{2})},\ \frac{-1}{n(n-1)(n-\frac{1}{2})}.$$

From these, by adjoining two copies each of 1 and -1, we can get a set of numbers for $k = 8, 9, 10, 11$ and by continuing in this way we arrive at the result for all $k > 3$.

REFERENCES

[1] J. W. S. Cassels, On a Diophantine equation, *Acta Arith.* 6 (1960) 47–52.

[2] G. Sansone and J. W. S. Cassels, Sur le problème de M. Werner Mnich, *Acta Arith.* 7 (1961/62) 187–190.

[3] H. E. Bohigian, Extensions of the W.Mnich problem, *Fib. Quarterly* 17 (1979) 172–77.

[4] È. T. Avanesov, Remarks on the problem of W. Mnich (Russian), *Mat. Časopis Sloven. Akad. Vied* 17 (1967) 85–91.

3 Polynomials Sums of Squares

In 1893 Hilbert showed that every semi-definite function in $R(x, y)$ is a sum of squares of elements in $R(x, y)$. In 1906 Landau showed that, in fact, no more than four squares were ever necessary.

Cassels, Ellison, and Pfister showed, in 1971, that the polynomial

$$f(x, y) = 1 + x^2(x^2 - 3)y^2 + x^2y^4$$

cannot be represented as a sum of three squares in $R(x, y)$.

This polynomial had been given by Motzkin in an entirely different context—see the first reference below.

A not quite related result is the following observation by Schinzel.

If four integral polynomials, not all constant, each of degree ≤ 4, is such that the sum of their cubes is a linear polynomial, then the constant in that linear polynomial is not congruent to 4 modulo 9.

REFERENCES

[1] J. W. S. Cassels, W. J. Ellison, and A. Pfister, On sums of squares and on elliptic curves over finite fields, *J. Number Theory* 3 (1971) 125–149.

[2] A. Schinzel, On sums of four cubes of polynomials, *J. London Math. Soc.* 43 (1968) 143–45.

Integer 4

1 Davenport-Baker Theorem

Diophantus (3rd century A.D.?) asked for four rational numbers such that the product of any two of them is one less than a rational square. He solved the problem finding x, $x + 2$, $4x + 4$, $9x + 6$ for $x = \frac{1}{16}$.

According to Dickson, Fermat (17th century) took 1, 3, 8 as three such numbers and found 120 as a fourth.

Euler was unable to find a fifth integer m such that the same would be true for 1, 3, 8, 120, m.

Van Lint (in the 1960s) asked for integers n such that the product of any two of $1, 3, 8, n$ was one less than a perfect square. In $3\frac{1}{2}$ minutes of computing time on an IBM 1620, it was shown that no such number n exists in the interval $120 < n < 10^{200}$.

In 7 minutes computing time, this was extended to show that 200 could be replaced by $1.7(10^6)$.

Using Baker's results on effective bounds for solutions to Diophantine equations van Lint and Baker showed (in March 1968 at Oberwohlfach) that no such n exists for $n > 10^{2(10^{487})}$.

The next year the gap was filled by Davenport and Baker to show that n must indeed be 120.

Veluppillai has given a very elementary argument showing that the set of numbers $\{2, 4, 12, 420\}$ also has the property that the product of each pair augmented by 1 is a square and such that no fifth number may be adjoined if this property is to be preserved. In fact, no other integer may replace 420.

It is not known if there is some set of 5 nonzero integers with the property that the product of any two of them is one less than a perfect square. Euler (see Dickson, p. 517) observed that when $ab + 1$ is a square, say c^2, then the four numbers

$$a,\ b,\ a + b + 2c,\ 4c(a + c)(b + c)$$

have the requisite property. Taking (a, b), respectively, to be $(1, 3)$; $(3, 8)$; $(2, 4)$; and $(8, 21)$ yields the quadruples $(1, 3, 8, 120)$; $(3, 8, 21, 2080)$; $(2, 4, 12, 420)$; and $(8, 21, 55, 37128)$.

If we take the a, b in Euler's prescription, given above, to be u_{2n}, u_{2n+2}, respectively, where u_n is the nth Fibonacci number ($u_1 = u_2 = 1$), then $ab + 1$ is the square of u_{2n+1}. Taking $c = u_{2n+1}$, Euler's quadruple turns out to be

$$u_{2n},\ u_{2n+2},\ u_{2n+4},\ 4u_{2n+1}u_{2n+2}u_{2n+3}.$$

This was shown, independently of the Euler result by Hoggatt and Bergum (see references below).

If one relaxes the restriction that the numbers be integers, but insists that they be rational, Euler gave (see Dickson, p. 517) the 5 numbers $1, 3, 8, 120, \frac{777480}{2879^2}$ whose products, in pairs, plus 1 are the squares of

$$2,\ 3,\ 11,\ 5,\ 19,\ 31,\ \frac{3011}{2879},\ \frac{3259}{2879},\ \frac{3809}{2879},\ \frac{10079}{2879}.$$

On p. 612 Dickson gives a method for finding sets of four rational numbers whose products, in triples, plus 1 are squares. A simple instance is the quadruple

$$\frac{5}{2},\ \frac{640}{81},\ 1,\ \frac{9}{40}.$$

It might be noted that the first three of x, $x + 2$, $4x + 4$, $9x + 6$ have the requisite property as polynomials. No fourth polynomial exists that will work with these three. However, Jones found all polynomials that work with x and $x + 2$.

He defined

$$f_k(x) = \frac{\alpha^k - \beta^k}{\alpha - \beta} \qquad \text{and} \qquad c_k(x) = 2f_{k(x)}f_{k+1}(x),$$

where α $(= \alpha(x))$ and β $(= \beta(x))$ are the roots of $w^2-2(x+1)w+1 = 0$. He then shows that

$$x,\ x+2,\ c_k(x),\ c_{k+1}(x)$$

are such that the product of any two of them is one less than a square.

Taking $x = k = 1$, we find $\alpha = 2 + \sqrt{3}$, $\beta = 2 - \sqrt{3}$, $f_1(1) = 1$, $f_2(1) = 4$, $f_3(1) = 15$, $c_1(1) = 8$, $c_2(1) = 120$. Thus, in this case, the quadruple x, $x + 2$, $c_1(1)$, $c_2(1)$ is just the quadruple 1, 3, 8, 120.

A somewhat similar problem was treated by Thamotherampillai who showed that the product of any two of $\{1, 2, 7\}$ plus 2 is a perfect square but the same is not true for any quadruple $\{1, 2, 7, c\}$, where $c > 1$.

Zheng showed that if the product of each pair of either $1, 2, 5, N$ or $1, 5, 10, N$ *diminished* by 1 is always a square, then $N = 1$. He does this by showing that each of the following two systems of equations is not possible in integers unless $x = 0$.

$$y^2 - 2x^2 = 1, \qquad z^2 - 5x^2 = 4,$$

$$y^2 - 5x^2 = 4, \qquad z^2 - 10x^2 = 9.$$

The deduction of the stated result from these systems is straightforward.

A related problem having to do with sums was investigated by J. Lagrange who showed that for $n \leq 5$ there are n integers such that they have pairwise square sums. For $n = 5$ the least solution is

$$(-4878,\ 4978,\ 6903,\ 12978,\ 31122).$$

In this case the sums, two at a time, are the squares of 10, 45, 90, 109, 134, 141, 162, 190, 195, 210.

Van Lint noted that for $N = 120$ each of the numbers $N + 1, 3N + 1, 8N + 1$ is a perfect square but that for no N in the range $120 < N < 10^{1700000}$ is the same thing true.

REFERENCES

[1] H. Davenport and A. Baker, The equations $3x^2-2 = y^2$ and $8x^2-7 = z^2$, *Quarterly J. of Math.* 20 (1969) 129–137.

[2] M. Veluppillai, The equations $z^2 - 3y^2 = -2$ and $z^2 - 6x^2 = -5$, in *A collection of manuscripts related to the Fibonacci sequence,* V. E. Hoggatt, Jr. and M. Bicknell-Johnson (eds.), Fibonnaci Association, 1980, pp. 71–75.

[3] V. E. Hoggatt and G. E. Bergum, A problem of Fermat and the Fibonacci sequence, *Fib. Quarterly* 15 (1977) 323–330.

[4] L. E. Dickson, *History of the Theory of Numbers*, vol. 2, Chelsea, New York, 1952.

[5] B. W. Jones, A variation on a problem of Davenport and Diophantus, *Quarterly J. of Math.* 27 (1976) 349–353.

[6] ——, A second variation on a problem of Diophantus, *Fib. Quarterly* 16 (1978) 155–165.

[7] N. Thamotherampillai, The set of numbers $\{1, 2, 7\}$, *Bull. Calcutta Math. Soc.* 72 (1980) 195–97.

[8] De Xun Zheng, On the systems of Diophantine equations $y^2 - 2x^2 = 1$, $z^2 - 5x^2 = 4$ and $y^2 - 5x^2 = 4$, $z^2 - 10x^2 = 9$ (Chinese), *Sichuan Daxue Xuebao* 24 (1987) 25–9.

[9] J. Lagrange, Cinq nombres dont les sommes deux á deux des carré, Sém. Delange-Pisot-Poitou, Exp. No. 20, 1970/71, 12 pp.

[10] P. Heichelheim, The study of positive integers (a, b) such that $ab + 1$ is a square, *Fib. Quarterly* 17 (1979) 269–274.

[11] J. Arkin, V. E. Hoggatt Jr., and E. G. Straus, On Euler's solution of a problem of Diophantus, *Fib. Quarterly* 17 (1979) 333–339.

[12] J. Arkin and G. Bergum, More on the problem of Diophantus, *Applications of Fibonacci Numbers,* (San Jose, 1986), Kluwer Acad. Publ., 1988, pp. 177–181.

[13] J. H. van Lint, On a set of Diophantine equations, Eindhoven Tech. Univ. T. H.-Report 68-WSK-03.

2 Arithmetic Progressions in the Fibonacci Sequence

There do not exist any arithmetic sequences of length 4 or more in the Fibonacci sequence. The only such sequences of length 3 are those of the form $u_{\ell-2}, u_\ell, u_{\ell+1}$ for $\ell > 2$.

REFERENCES

W. Sierpiński, *250 Problems in Elementary Number Theory*, American Elsevier, New York, 1970, #64.

3 Factorials

There are exactly 4 integers that are equal to the sum of the factorials of their base 10 digits and they are 1, 2, 145, 40585.

That the number of such integers is finite is a consequence of a general theorem due to Schwartz [2]. Let A be a natural number and let its base b digits, from units up, be $a_0, a_1, \ldots, a_n$. Further, let $\{f_n(\cdot)\}$ be a family of functions each defined, at least, on the first b nonnegative integers. One calls *A self-generating with respect to the f_n* if $A = S(A)$, where $S(A) = \sum_{i=0}^{n} f_n(a_i)$. With this notation Schwartz's theorem reads as follows. *If*

$$\limsup_n \frac{n \max_{0 \le i \le b-1} f_n(i)}{b^n} < 1,$$

then the set of all self-generating integers with respect to the f_n is finite.

If we take $f_n(x) = x!$ for all n, the theorem tells us there are only finitely many integers equal to the sum of the factorials of their digits.

If we take $f_n(x) = x^n$ for all n, the theorem tells us there are only finitely many integers, having n digits, equal to the sum of the nth powers of their digits. Elsewhere he shows that 10^{60} is an upper bound for such a number. (An example is $153 = 1^3 + 5^3 + 3^3$. A non-example, since the number of digits, 8, in the number is not the same as the right-hand exponent, 7, is

$$14459929 = 1^7 + 4^7 + 4^7 + 5^7 + 9^7 + 9^7 + 2^7 + 9^7.)$$

In his paper, "Finiteness of a Set of Self-Generating Integers," Schwartz listed all of the known (to that time) such numbers up to 10 digits. For $n = 4$ he gives 1634, 8208, 9474 and for $n = 10$ he lists only 4679307774. At the time no larger instances were known.

In the solution to an elementary problem in *The American Mathematical Monthly,* N. J. Fine showed that no base 10 representation of an integer has the property that the sum of the squares of its digits add to the number. He also showed that if the base b of representation is such that $b^2 + 1$ is composite, then, and only then, can the sum of the squares of the digits add to the number.

Subramanian proved that in a given base b if the sum of the cubes of the digits of an integer is equal to the integer, then the integer may not have more than 4 digits. Further, if the last digit is either 0 or 1, then it cannot have more than 3 digits and if $\sqrt{b}$ is irrational, then it must have exactly three digits.

Questions of this kind have gained a certain kind of notoriety from the following statement of Hardy appearing in his *A Mathematician's Apology*.

> There are just four numbers (after 1) which are the sums of the cubes of their digits, viz.
>
> $$153 = 1^3 + 5^3 + 3^3, \qquad 370 = 3^3 + 7^3 + 0^3,$$
>
> $$371 = 3^3 + 7^3 + 1^3, \qquad 407 = 4^3 + 0^3 + 7^3.$$
>
> These are odd facts, very suitable for puzzle columns and likely to amuse amateurs, but there is nothing in them which appeals to a mathematician. The proofs are neither difficult nor interesting—merely a little tiresome. The theorems are not serious; and it is plain that one reason (though perhaps not the most important) is the extreme specialty of both the

> enunciations and the proofs, which are not capable of any significant generalization.

Hardy does try, elsewhere in his book, to elucidate what he means by a "serious theorem" and a "significant result."

Hardy's first example of a nonserious theorem is that the only integers less than 10000 which are multiples of their 'reversals' are $8712 = 4 \cdot 2178$, $9801 = 9 \cdot 1089$. In a recent remark in *The Mathematical Gazette,* Beech said, "... my Apple IIc confirmed that for all integers up to 10000 the only numbers which are multiples of their reversals are 8712 and 9801 ... a few more hours of running time were required to test all integers up to 100000 ... Between 10000 and 100000 the only integers which are integral multiples of their reversals are 87912 and 98901 with $87912 = 4 \cdot 21978$ and $98901 = 9 \cdot 10989$."

The obvious generalization holds where the single central 9 is replaced by a string of 9's. Thus the "computer conjecture" is that this exhausts all possible such numbers.

Comments similar to Hardy's, in a somewhat more "serious" context were made by Hirzebruch in an article titled "The signature theorem: Reminiscences and recreation" (pp. 3–4).

> The old signature theorem involves Bernoulli numbers, and has many relations to number theory and applications in topology. Exotic spheres were discovered using it. ... The equivariant signature theorem has many more number theoretical connections. In the second half of this lecture we shall point out some rather elementary connections to number theory obtained by studying the equivariant signature theorem for four-dimensional manifolds. Perhaps these connections still belong to recreational mathematics because no deeper explanations, for example of the occurrences of Dedekind sums both in the theory of modular forms and in the study of four-dimensional manifolds, is known.

REFERENCES

[1] G. D. Poole, Integers and the sum of the factorials of their digits, *Math.*

Mag. 44 (1971) 278–89.

[2] B. L. Schwartz, Self-generating integers, *Math. Mag.* 46 (1973) 158–160.

[3] ——, Finiteness of a set of self-generating integers, *J. Rec. Math.* 2 (1969) 79–83.

[4] N. J. Fine, Elem. Prob. E1651, *Amer. Math. Monthly* 71 (1964) 90, 1042–43.

[5] P. K. Subramanian, On bases and cycles, *Math. Mag.* 41 (1968) 117–123.

[6] G. H. Hardy, *A Mathematician's Apology*, Cambridge U. Press, Cambridge, 1967.

[7] M. Beech, A computer conjecture of a non-serious theorem, *Math. Gaz.* 74 (1990) 50–51.

[8] F. Hirzebruch, The signature theorem: Reminiscences and recreation, in *Prospects in Mathematics*, F. Hirzebruch, L. Hormander, J. Milnor, J.-P. Serre, and I. M. Singer, Annals of Mathematics Studies #70, Princeton U. Press, Princeton, 1971, pp. 3–41.

4 The Next to Last Case of a Factorial Diophantine Equation

Pollack and Shapiro considered the equation $n! + 1 = x^m$. They observed:

1. for $m = 3$ or $m > 4$ there are no solutions;
2. for $m = 2$ and $n = 4, 5, 7$ there are solutions;
3. for $m = 4$ there are no solutions.

In their paper the last of these is proved, the others having been proved before. The proof proceeds in two stages. First, it is shown there is no solution for $n > 27182$ and then a computer search shows there are none less than or equal to this integer. It is not known if the set of solutions given above contains all solutions.

REFERENCES

R. M. Pollack and H. N. Shapiro, The next to last case of a factorial Diophantine equation, *Comm. Pure Appl. Math.* 26 (1973) 313–325.

5 Hall's Theorem

Every real number has a simple continued fraction expansion. Except in the case of rational numbers these expansions are unique, and, even in the case of rational numbers, there are only two such expansions, exactly one of which will have as last nonzero partial quotient the integer 1. When we speak of "the" continued fraction expansion of a real number we will, when there is an ambiguity, be referring to this expansion with last partial quotient 1.

Let $[a_1, a_2, a_3, \ldots]$ be the simple continued fraction expansion of the real number a. Define

$$F(n) = \{a | a_i \leq n \text{ for all } i \geq 1\}.$$

The set of numbers $F(4)$ is of special interest. A theorem of Marshall Hall reads as follows.

i. *Every real number a may be written $a = b + c$, where b and c are in $F(4)$.*

ii. *Every real number a may be written $a = bc$, where b and c are in $F(4)$.*

iii. *If $\frac{3}{2} = x + y$ or $4 = xy$, then either infinitely many x_i are ≥ 4 or infinitely many y_i are ≥ 4 or infinitely many of the x_i and infinitely many of the y_i are equal to 3.*

Cusick and Lee let $S(k)$ be the set of numbers between 0 and $\frac{1}{k}$ whose simple continued fraction partial quotients are all smaller than k and showed that each real number in the unit interval may be written as a sum of k elements of $S(k)$.

Diviš and Cusick, independently, showed that every real number is in the sum of four copies of $F(2)$ and is in the sum of three copies of $F(3)$. In each case fewer copies will not suffice.

Further results, due to Hlavka, are the following:

$$F(3) + F(4), \qquad F(2) + F(2) + F(4),$$
$$F(2) + F(7), \qquad F(2) + F(3) + F(3)$$

are all equal to the set of all real numbers and

$$F(2) + F(4), \qquad F(3) + F(3), \qquad F(2) + F(2) + F(3)$$

are not equal to the set of all real numbers.

REFERENCES

[1] M. Hall, On the sum and product of continued fractions, *Annals of Math.* 48 (1947) 966–993.

[2] B. Diviš, On sums of continued fractions, *Acta Arith.* 22 (1973) 157–173.

[3] T. W. Cusick and R. A. Lee, Sums of sets of continued fractions, *Proc. Amer. Math. Soc.* 30 (1971) 241–46.

[4] T. W. Cusick, On M. Hall's continued fraction theorem, *Proc. Amer. Math. Soc.* 38 (1973) 253–54.

[5] J. L. Hlavka, Results on sums of continued fractions, *Trans. Amer. Math. Soc.* 211 (1975) 123–134.

6 Residue Classes

Let $r_1, r_2, \ldots, r_k$ be distinct residue classes modulo m. Does there necessarily exist an integral polynomial f such that

$$f(x) \equiv 0 \pmod{m}$$

has precisely $r_1, \ldots, r_k$ as solutions? The answer is yes if

$$m = 4 \text{ or } m \text{ is an odd prime}$$

and the answer is no if

$$m > 4 \text{ and } m \text{ is composite.}$$

REFERENCES

M. M. Chojnacka-Pniewska, Sur les congruences aux racines données, *Ann. Polon. Math.* 3 (1956) 9–12.

7 Diophantine Representation

There is a well-known theorem, attributed to Lagrange, that states that every integer may be written as a sum of four squares. Among other things, Matijasevič has shown that every integer has a representation in the form $a^2 + b^2 + c^2 + c + 1$.

See also Integer 5, Primes as Range of a Polynomial.

REFERENCES

Ju. V. Matijasevič, A Diophantine representation of the set of prime numbers (Russian), *Dokl. Akad. Nauk SSSR* 196 (1971) 770–773.

8 Lattice Cornered Polygons

A regular polygon in the plane with lattice point vertices must be a square.

If a triangle in the plane with lattice point vertices contains exactly one lattice point in its interior, then it must have one of 3, 4, 6, 8, or 9 lattice points on its boundary.

Pick's theorem says that $A = \frac{1}{2}B + I - 1$, where A, B, I are, respectively, the area and the numbers of lattice points on the boundary and in the interior of a simple closed polygon whose vertices are lattice points.

In an interesting paper, DeTemple and Robertson show the equivalence of Pick's theorem and Euler's theorem (relating the numbers of vertices, faces, and edges of a plane figure).

As Reeve pointed out in 1957 there can be no three-dimensional exact analogue of Pick's theorem since all of the tetrahedra in three space with vertices $(0,0,0)$, $(1,0,0)$, $(0,1,0)$, and $(1,1,r)$, where r is a positive integer, have the same number of lattice points inside and on their edges and faces but their volumes tend to infinity as r increases.

He called the set of ordinary lattice points *the fundamental lattice* and denoted it by L. He then introduced a new lattice, denoted by L_n, for each positive integer n. The point (a,b,c) is in L_n if and only if the point (na,nb,nc) is in L. Letting $L_n(\Gamma), L(\Gamma), L_n(\overline{\Gamma}), L(\overline{\Gamma})$ be the respective numbers of points of L_n, L in the polyhedron Γ and on its boundary $\overline{\Gamma}$, the following connection of these quantities and the volume of the polyhedron holds. For $n > 1$

$$2n(n-1)(n+1)V(\Gamma) = 2\{L_n(\Gamma) - nL(\Gamma)\} - \{L_n(\overline{\Gamma}) - nL(\overline{\Gamma})\}.$$

Reeve observed, "We feel that the interest of our results lies in some respects not so much in the form of the formula we obtain for the volume of a polyhedron as in the fact that the introduction of a secondary lattice enables us to obtain such a formula at all."

A recent paper by Scott in the *American Mathematical Monthly* briefly discusses Pick's theorem and the results of Reeve.

See also Integer 8, Regular Lattice n-Simplices.

REFERENCES

[1] D. G. Ball, The constructibility of regular and equilateral polygons on a square pinboard, *Math. Gaz.* 57 (1973) 119–122.

[2] C. S. Weaver, Geoboard triangles with one interior point, *Math. Mag.* 50 (1977) 92–94.

[3] D. DeTemple and J. M. Robertson, The equivalence of Euler's and Pick's theorems, *Math. Teacher* 67 (1974) 222–26.

[4] J. E. Reeve, On the volume of lattice polyhedra, *Proc. London Math. Soc.* 7 (1957) 378–395.

[5] P. R. Scott, The fascination of the elementary, *Amer. Math. Monthly* 94 (1987) 759–768.

9 von Neumann Algebras

Finite-dimensional von Neumann algebras may have any "index" > 4 but only certain indices between 1 and 4. (Vaughan Jones)

REFERENCES

Notices of the Amer. Math. Soc. 32 (1985) 352.

10 A Few Diophantine Equations

There are exactly 4 positive integer pairs x, y for which

$$1 + 2 + \cdots + y = 1^2 + 2^2 + \cdots + x^2$$

and they are

$$(x, y) = (1, 1), (5, 10), (6, 13), (85, 645).$$

Using the expressions for the sum of the first y integers and the sum of the first x squares of integers, the above equation is equivalent, after multiplying by 6, to

$$3y(y + 1) = x(x + 1)(2x + 1).$$

Similarly, Cohn has shown the Diophantine equation

$$y(y+1)(y+2)(y+3) = 2x(x+1)(x+2)(x+3) \qquad (1)$$

has exactly one solution and it is $(4,5)$.

Using techniques of Cohn the equation obtained from (1) by replacing the factor 2 by 3 was shown by Ponnudurai to have only the solutions $(2,3)$ and $(5,7)$.

Boyd and Kisilevsky have shown the equation

$$y(y+1)(y+2)(y+3) = x(x+1)(x+2)$$

has only the solutions

$$(x,y) = (2,1), (4,2), (55,19).$$

The equation $x(x+1)(x+2) = y(y+1)(2y+1)$ was shown by Finkelstein [6] to have only the solutions $(x,y) = (0,0)$ and $(1,1)$.

The equation $x(x+1)(x+2) = y(y+1)$ was shown by Mordell to have solutions only for $x = -2, -1, 0, 1, 5$.

The remainder of this section is primarily devoted to the so-called *Catalan* equation

$$x^p - y^q = 1, \qquad p > 1,\ q > 1,\ x > 1,\ y > 1.$$

(Throughout the rest of the section these inequalities on p, q, x, y are assumed.)

Catalan (in 1844) conjectured that $3^2 - 2^3 = 1$ was the only solution of this equation.

A long time before (in 1738) Euler had shown that this solution is the only solution of the equation $x^2 - y^3 = 1$.

Lebesgue (in 1850) proved $x^p - y^2 = 1, p \neq 3$ has no solution.

Nagell (in 1921) showed that neither $x^3 - y^q = 1$ nor $x^p - y^3 = 1, p \neq 2$, has a solution.

Selberg (in 1932) showed $x^4 - y^3 = 1$ has no solution.

Chao Ko (in 1964) showed $x^2 - y^q = 1$ has no solution. (A very simple proof of this when q is a prime larger than 3 was given by Chein in 1976.)

Cassels (in 1960) showed that if $x^p - y^q = 1$ has a solution, then $p|y$ and $q|x$. From this he deduced that there do not exist 3 consecutive integers which are all perfect powers.

Tijdeman (in 1976) showed $x^p - y^q = 1$ has only finitely many solutions.

Tijdeman's paper, "On the Equation of Catalan," has a nice historical introduction and a good bibliography. For an even more complete exposition of the problem see the paper by Ribenboim.

Gabard discussed the equation $y^2 - 18x^2 = 1$, whose solutions are given by $y_{n+2} = 34y_{n+1} - y_n$, and observed that Thébault showed the solutions y_{2m+1}, $m > 0$ are all composite. Gabard then showed that

$$y_1 = 17, \qquad y_2 = 577, \qquad y_4 = 665837$$

are prime and that y_8, y_{16} are composite.

For a recent account of the Diophantine equation $y^2 - x^3 = k$ see pp. 233–240 of Tijdeman [12].

See also Integer 3, Sums; Integer 5, Figurate Numbers; Integer 17, The Equation $x^3 - y^2 = -17$; and Integer 24, Sum of Consecutive Squares a Square.

REFERENCES

[1] É. T. Avanesov, The Diophantine equation $3y(y+1) = x(x+1)(2x+1)$, *Volz. Mat. Sb. Vyp.* 8 (1971)3–6.

[2] R. Finkelstein and H. London, On triangular numbers which are sums of consecutive squares, *J. Number Theory* 4 (1972) 455–462.

[3] J. H. E. Cohn, The Diophantine equation $y(y+1)(y+2)(y+3) = 2x(x+1)(x+2)(x+3)$, *Pac. J. Math.* 37 (1971) 331–35.

[4] T. Ponnudurai, The Diophantine equation $Y(Y+1)(Y+2)(Y+3) = 3X(X+1)(X+2)(X+3)$, *J. London Math. Soc.* 10 (1975) 232–240.

[5] J. W. Boyd and H. H. Kisilevsky, The Diophantine equation $u(u+1)(u+2)(u+3) = v(v+1)(v+2)$, *Pac. J. Math.* 40 (1972) 23–32.

[6] R. Finkelstein, On a Diophantine equation with no nontrivial integral solution, *Amer. Math. Monthly* 73 (1966) 471–77.

[7] L. J. Mordell, On the integer solutions of $y(y+1) = x(x+1)(x+2)$, *Pac. J. Math.* 13 (1963) 1347–1351.

[8] E. Z. Chein, A note on the equation $x^2 = y^q + 1$, *Proc. Amer. Math. Soc.* 56 (1976) 83–4.

[9] R. Tijdeman, On the equation of Catalan, *Acta Arith.* 29 (1976) 197–209.

[10] P. Ribenboim, Consecutive powers, *Exp. Math.* 2 (1984) 193–221.

[11] E. Gabard, Factorisations et équation de Pell, *Mathesis* 67 (1958) 218–220.

[12] R. Tijdeman, Diophantine equations and Diophantine approximations, in *Number Theory and Applications* (ed. R. A. Mollin) Kluwer, Dordrecht, 1988, pp. 215–243.

II Four Fifth Powers

Euler made the conjecture, in 1769, that no kth power of an integer is the sum of fewer than k kth powers of integers. This is false since, as shown by Lander and Parkin in 1967,

$$27^5 + 84^5 + 110^5 + 133^5 = 144^5.$$

When Lander and Parkin carried out their search no examples of the sum of three integer fourth powers equal to an integral fourth power came to light.

In 1988 Elkies found infinitely many such sums of fourth powers. The first example he found was

$$2682440^4 + 15365639^4 + 18796760^4 = 20615673^4.$$

Though this was the first one he found it is not the smallest. Using his work R. Frye (see Elkies below) found the smallest to be:

$$95800^4 + 217519^4 + 414560^4 = 422481^4.$$

REFERENCES

[1] L. J. Lander and T. R. Parkin, A counterexample to Euler's sum of powers conjecture, *Math. Comp.* 21 (1967) 101–103.

[2] N. D. Elkies, On $A^4 + B^4 + C^4 = D^4$, *Math. Comp.* 51 (1988) 825–835.

12 Stufe of a Real Quadratic Field

One defines the *stufe* of a quadratic field k to be the least positive integer n, if there is one, for which -1 is a sum of n squares over k. If no such sum exists, then the stufe is ∞. The stufe of a real quadratic field $Q(\sqrt{-d})$ is always either 1, 2, or 4 according to whether $d = 1$, $1 \neq d \not\equiv -1 \pmod 8$, or $d \equiv -1 \pmod 8$.

REFERENCES

[1] T. Nagell, Sur la résolubilité de l'equation $x^2 + y^2 + z^2 = 0$ dans un corps quadratique, *Acta Arith.* 21 (1972) 35–43.

[2] K. Szymiczek, Note on a paper by Nagell, *Acta Arith.* 25 (1974) 313–14.

Integers 4, 5

1 A Problem of Znám

In 1972 Znám asked if for each positive integer n there were integers x_i, $1 \le i \le n$, such that each of the x_i was a proper divisor of the product of the others plus 1.

In 1975 Skula showed this not to be the case for $2 \le n \le 4$.

Janák gave the set 2, 3, 11, 23, 31 meeting the conditions when $n = 5$.

In 1978 Janák and Skula gave all solutions in the cases $n = 5, 6$, and also gave 18 solutions in the case $n = 7$.

In 1987 Cao Zhenfu, Liu Rui, and Zhang Liangrui showed that in the case $n = 7$ there are exactly 23 solutions and they tabulated them all. Further, they made the conjecture, true for $1 \le s \le 6$, that the congruence, in which all p_j are primes,

$$2p_1 \cdots p_{j-1}p_{j+1} \cdots p_s + 1 \equiv 0 \pmod{p_j}, \qquad 1 \le j \le s,$$

has exactly one solution. In the case $s = 6$, they gave the one solution as

$$p_1 = 3,\ p_2 = 11,\ p_3 = 17,\ p_4 = 101,\ p_5 = 149,\ p_6 = 3109.$$

They also stated that Znám's problem had been completely solved by Sun Qi (On a problem of Znám, *Sichuan Daxue Xuebao* 4(1983) 9–11).

REFERENCES

[1] L. Skula, On a problem of Znám, *Acta Fac. Rerum Natur. Univ. Comenian Math.* 32 (1975) 87–90.

[2] J. Janák and L. Skula, On the integers x_i for which $x_i | x_1 \cdots x_{i-1} x_{i+1} \cdots x_n + 1$ holds, *Math. Slovaca* 28 (1978) 305–310.

[3] Cao Zhenfu, Liu Rui and Zhang Liangrui, On the equation $\sum_{j=1}^{s} \frac{1}{x_j} + \frac{1}{x_1 \cdots x_s} = 1$ and Znám's problem, *J. Number Theory* 27 (1987) 206–211.

2 An Awful Problem

Loxton and van der Poorten state the following: "The set Z of all integers coincides with the language of all words on the symbols $0, 1, \overline{1}$, and 2 interpreted as integers in base 4; here $\overline{1}$ is a convenient contraction for the digit -1. We consider the subset L of Z omitting the digit 2; thus the language of all words on just the symbols 0, 1 and $\overline{1}$ interpreted as integers in base 4. Our problem is this: can every odd integer be written as a quotient of elements of L?"

We are told the question arose in connection with a paper by Brown, Moran, and Tijdeman titled "Riesz products are basic measures."

We are also told "The problem became known as 'that awful problem about integers in base 4' to those unfortunate enough to have become obsessed or intrigued by it."

The answer turns out to be 'yes.' However, the same problem for bases larger than 4 turns out to have a negative answer. It is not known if the answer is yes or no in the case of base 3.

A related problem, and one referred to in the above cited paper, was treated by Lehmer, Mahler, and van der Poorten. Let g be an integer greater than or equal to 2 and let L be the set of nonnegative integers whose base g representations use only the digits 0 and 1. For a given k the authors study the congruence

$$\ell \equiv a \pmod{k}, \qquad \text{for } \ell \epsilon L.$$

This congruence is either solvable with an infinity of solutions or it is not solvable. In fact, for given g and $k \geq 1$ let L_a be the set of ℓ satisfying the above congruence. Defining m and D to be 1 when g and k are relatively prime and otherwise $D = (g^m, k)$, where m is the least positive integer such that $(g^m, k) = (g^{m+1}, k)$, it is shown that L_a is infinite if and only if there exists an ℓ of m digits satisfying the congruence $\ell \equiv a \pmod{D}$. Otherwise L_a is empty.

REFERENCES

[1] J. H. Loxton and A. J. van der Poorten, An awful problem about integers in base 4, *Acta Arith.* 49 (1987) 193–203.

[2] D. H. Lehmer, K. Mahler, and A. J. van der Poorten, Integers with digits 0 or 1, *Math. Comp.* 46 (1986) 683–89.

Integer 5

1 Euler Polynomials

Define the Euler and the Bernoulli polynomials $E_n(x)$ and $B_n(x)$ by the following equations:

$$\sum_{n=0}^{\infty} \frac{E_n(x)z^n}{n!} = \frac{2e^{xz}}{e^z + 1} \qquad \text{and} \qquad \sum_{n=0}^{\infty} \frac{B_n(x)z^n}{n!} = \frac{ze^{xz}}{e^z - 1}.$$

The only Euler polynomial that has a multiple root is $E_5(x)$. The Bernoulli polynomial B_{2n+1} is divisible by $x^2 - x - 1$ if and only if $n = 5$.

The factorization of B_{11} is given by:

$$3B_{11}(x) = x\left(x - \frac{1}{2}\right)(x^2 - x - 1)(3x^6 - 9x^5 + 2x^4 + 11x^3 + 3x^2 - 10x - 5).$$

If one defines the *Bernoulli numbers* B_n by the equation $B_n = B_n(0)$, then an amusing congruence proved by Frame is

$$-30B_{4m} \equiv 1 + 6000\binom{m-1}{2} \pmod{27000}.$$

REFERENCES

[1] J. Brillhart, On the Euler and Bernoulli polynomials, *J. für Reine und Angew. Math.* 234 (1969) 45–64.

[2] J. S. Frame, Bernoulli numbers modulo 27000, *Amer. Math. Monthly* 68 (1961) 87–95.

2 Primes a Sum and a Difference of Primes

The integer 5 is the only prime that is the sum and the difference of two primes.

(It is not hard to see that if p is such a prime, then $p-2, p, p+2$ must all be prime. But this is not possible unless $p = 5$.)

Cohen and Selfridge have proved that not every integer is a sum or difference of two prime powers. They showed the existence of infinitely many odd M such that for all n neither $M+2^n$ nor $|M-2^n|$ is a power of a prime. If every integer were a sum or difference of powers of primes, then such an M would have to be of the form $2^s + p^t$ for some integers s, t and prime p. But then this integer minus 2^s would be a power of a prime. In their paper, they gave an example of a 94 digit instance of such an M.

Earlier, Sierpiński observed the existence of infinitely many even integers each of which is a sum and a difference of two primes and the existence of infinitely many odd integers which are neither a sum nor a difference of two primes.

REFERENCES

[1] F. Cohen and J. L. Selfridge, Not every integer is the sum or difference of two prime powers, *Math. Comp.* 29 (1975) 79–81.

[2] W. Sierpiński, Sur les nombres qui sont sommes et différences de deux nombres premiers, *U. Beograd Publ. Electrotehn. Fak. Ser. Mat. Fiz.*, No. 84-91 (1963) 1–2.

3 Figurate Numbers

Sierpiński, in his book *A Selection of Problems in the Theory of Numbers*, observed that Escott and Sendacka have shown that the only numbers less than 10^9 which are both triangular, of the form $\binom{n}{2}$, and tetrahedral, of the form $\binom{m}{3}$, are the five numbers

$$1,\ 10,\ 120,\ 1540,\ 7140$$

The pairs of values for n, m in these five numbers are $(2,3)$, $(5,5)$, $(16,10)$, $(56,22)$, $(120,36)$.

Avanesov proved that these five do indeed exhaust all possible numbers which are both triangular and tetrahedral.

Singmaster showed that three binomial coefficients $\binom{n}{k}$, $\binom{n}{k+1}$, $\binom{n}{k+2}$ can be in the ratio $1:2:3$ precisely for $n = 14$, $k = 4$ and this triple of binomial coefficients form the following portion of Pascal's triangle.

1001 2002 3003
3003 5005
8008

Tzanakis and de Weger showed that the equation $y^2 = x^3 - 4x + 1$ has the following 22 solutions

$$(x, \pm y) = (-2,1),\ (-1,2),\ (0,1),\ (2,1),\ (3,4),\ (4,7),\ (10,31),$$
$$(12,41),\ (20,89),\ (114,1217),\ (1274,45473).$$

From this they conclude that the only triangular numbers which are products of three consecutive integers are:

$$T_3,\ T_{15},\ T_{20},\ T_{44},\ T_{608},\ T_{22736}.$$

Mohanty had announced this exact same result the previous year. However, as was pointed out in a letter to the editor by A. Bremner, there were mistakes in Mohanty's arguments. Despite the fact that the main methods of the paper were not at all elementary, it is interesting to note that elementary errors did creep in. One of the errors, quoted by Bremner, is the following. In the paper, one finds the equation

$$12v^2 - 4uv - u^2 = \frac{(2v^2 - 1)(2v^2 + 1)}{u^2}$$

from this the author concludes "since $2v^2 - 1, 2v^2 + 1$ are two consecutive odd integers, u^2 divides one of them but not both." Bremner observes that, with $u = 21$ and $v = 142$, u^2 divides $(2v^2 - 1)(2v^2 + 1)$, but u^2 divides neither of the two factors.

Unfortunately, finding mistakes in published papers is not at all uncommon—even in refereed journals. It is not, however, always the case that the main results are unaffected by the fallacious arguments.

On p. 56 of *Elementary Theory of Numbers*, Sierpiński , answering a question of Zarankiewicz , gives an example of a Pythagorean triple made up of triangular numbers. The example is $(8778, 10296, 13530)$ and they are $T_{132}, T_{143}, T_{164}$, respectively.

As an easy exercise he leaves showing that all of the numbers $21, 2211, 222111, \ldots$ are triangular to the reader.

See also Integer 4, A Few Diophantine Equations.

REFERENCES

[1] È. T. Avanesov, Solution of a problem of figurate numbers (Russian), *Acta Arith.* 12 (1966/67) 409–420.

[2] D. Singmaster, Repeated binomial coefficients and Fibonacci numbers, *Fib. Quarterly* 13 (1975) 295–98.

[3] S. P. Mohanty, Integral points of $y^2 = x^2 - 4x + 1$, *J. Number Theory* 30 (1988) 86–93.

[4] N. Tzanakis and B. M. M. de Weger , On the practical solution of the Thue equation, *J. Number Theory* 31 (1989) 99–132.

[5] A. Bremner, Letter to the editor, *J. Number Theory* 31 (1989) 373.

[6] W. Sierpiński, *Elementary Theory of Numbers*, North-Holland, Amsterdam, 1988.

4 Consecutive Primitive Roots

The integers 2, 3, 4, and 6 each have exactly one primitive root and therefore, by default, each has a set of primitive roots consisting of "consecutive" integers.

The integer 5, with primitive roots of 2 and 3 is the only positive integer having at least two primitive roots for which the entire set of primitive roots are consecutive integers. This is a consequence of the following theorems, due to Monzingo. (Recall that only 2, 4, powers of odd primes, and twice such powers have primitive roots.)

Theorem 4.1. *If $m = 2p^n$ $(m > 6)$, $n \geq 1$, p an odd prime, then the primitive roots of m are not consecutive.*

Theorem 4.2. *If $m = p^n$, $n \geq 2$, p an odd prime, then the primitive roots of m are not consecutive.*

Lemma 4.3. *If p is an odd prime greater than 13 but not one of 19, 31, 43, or 61, then $2\sqrt{p-1} \leq \phi(p-1)$.*

Theorem 4.4. *If p is a prime $p > 5$, then the primitive roots of p are not consecutive.*

If one is interested in knowing just those primes containing some *pair* of consecutive primitive roots one has the following result of Cohen.

Every finite field of order other than 2, 3, 7 contains a pair of consecutive primitive roots.

A well-known conjecture of Artin says that every square-free integer larger than 1 is a primitive root of infinitely many primes. Very little toward settling this conjecture was accomplished until fairly recently.

In 1984 Gupta and Ram Murty proved the following result: *For q, r, s distinct primes at least one of the numbers*

$$qs^2,\ q^3r^2,\ q^2r,\ r^3s^2,\ r^2s,\ q^2s^3,\ qr^3,\ q^3rs^2,\ rs^3,\ q^2r^3s,\ q^3s,\ qr^2s^3,\ qrs$$

is a primitive root modulo p for infinitely many p.

In 1986 Heath-Brown proved that at least one of 2, 3, 5 is a primitive root for infinitely many primes.

Ram Murty wrote an interesting expository article on the Artin conjecture in the *Mathematical Intelligencer* in 1988. In his article, Murty gave an example that seems surprising at first glance but which, on reflection, makes us surprised at our being surprised. He observed that the binary expansion of $\frac{1}{98,007,599}$ has almost fifty million digits in its period. On reflection one notes that 2 is a primitive root of each of 9851 and 9949, whose product is 98,007,599, and, therefore, the fraction $\frac{1}{98,007,599}$ has a base 2 expansion with period length the least common multiple of 9850 and 9948—48,993,900.

REFERENCES

[1] M. G. Monzingo, On consecutive primitive roots, *Fib. Quarterly* 14 (1976) 391, 394.

[2] S. D. Cohen, Pairs of primitive roots, *Mathematika* 32 (1985) 276–285.

[3] R. Gupta and M. Ram Murty, A remark on Artin's conjecture, *Invent. Math.* 78 (1984) 127–130.

[4] D. R. Heath-Brown, Artin's conjecture for primitive roots, *Quarterly J. Math.* 37 (1986) 27–38.

[5] M. Ram Murty, Artin's conjecture for primitive roots, *Math. Intelligencer* 10 (4) (1988) 59–67.

5 Five Moduli

Only when m is one of the five numbers 1, 2, 6, 42, 1806 does the truth of the two congruences $a \equiv b \pmod{m}$, $c \equiv d \pmod{m}$, where a, b, c, d are positive integers, necessarily entail the truth of the congruence $a^c \equiv b^d \pmod{m}$.

REFERENCES

[1] J. Dyer-Bennet, A theorem on partitions of the set of positive integers, *Amer. Math. Monthly* 47 (1940) 152–54.

[2] W. Sierpiński, *Elementary Theory of Numbers* North-Holland, Amsterdam, 1988, pp. 275–76.

6 A Conjecture Based on Goldbach's Conjecture

Goldbach's conjecture, a very old conjecture in the theory of numbers, asserts that every even integer larger than 2 is the sum of two primes. Much progress has been made on the problem though it still remains unproved. For information on the current state of affairs see Sierpiński.

It is known that $\sigma(n) - n$ does not take on the values 2, 5, 52, 88 and that it does take on all other values from 1 to 51. Using Goldbach's conjecture, we see that since $\sigma(pq) - pq = 1 + p + q$, p, q odd primes, the truth of the conjecture implies that all odd numbers larger than 5 are values of the quantity in question.

Hence, if Goldbach's conjecture is true, the only *odd* integer not a value of $\sigma(n) - n$ is the integer 5.

Erdős has shown that there exist infinitely many positive integers not of the specified form.

Though, as we have observed, Goldbach's conjecture remains unproved it is interesting to note that the corresponding assertion for polynomials is easy to settle. In fact, Hayes has shown that an arbitrary polynomial of degree n may be written as the sum of two irreducible polynomials each of degree n. We show this in the following paragraph.

Let $f(x) = c_n x^n + \cdots + c_0$ be an integral polynomial of degree n. Choose different odd primes p, q such that neither of them divides $c_n c_0$. For each i, $0 \leq i \leq n$, there are integers s_i, t_i such that $s_i p + t_i q = c_i$. Now p can divide at most one of s_0, $s_0 + q$, $s_0 + 2q$ and q can divide at most one of t_0, $t_0 - p$, $t_0 - 2p$. Hence, for at least one value of ϵ equal to 0, 1, or 2, we have $p \nmid s_0 + \epsilon q$ and $q \nmid t_0 - \epsilon p$.

Putting

$$g(x) = (t_n q)x^n + s_{n-1} p x^{n-1} + \cdots + s_1 p x + (s_0 + \epsilon q)p$$

and

$$h(x) = (s_n p)x^n + t_{n-1} q x^{n-1} + \cdots + t_1 q x + (t_0 - \epsilon p)q$$

we see that $f(x) = g(x) + h(x)$ and, using the Eisenstein criterion twice, once with respect to p for $g(x)$ and once with respect to q for $h(x)$, that $g(x)$ and $h(x)$ are irreducible.

See Integer 3, Iterates.

REFERENCES

[1] L. Vojtech, On some problems in the elementary theory of numbers, *Acta Fac. Rerum Natur. Univ. Commen. Math.* 32 (1975) 47–67.

[2] P. Erdős, Über die Zahlen der Form $\sigma(n) - n$ und $n - \phi(n)$, *Elem. Math.* 28 (1973) 83–86.

[3] W. Sierpiński, *Elementary Theory of Numbers*, North-Holland, Amsterdam, 1988.

[4] D. R. Hayes, A Goldbach theorem for polynomials with integral coefficients, *Amer. Math. Monthly* 72 (1965) 45–46.

7 A Square the Sum of Three Squares

Given n a positive integer the system

$$n^2 = x^2 + y^2 + z^2, \qquad xyz \neq 0$$

may be solved for x, y, z if and only if n is neither a power of 2 nor 5 times such a power.

This is based on Lebesgue's identity

$$(a^2+b^2+c^2+d^2)^2 = (a^2+b^2-c^2-d^2)^2 + (2ac+2bd)^2 + (2ad-2bc)^2.$$

There is a large literature concerning sums of three squares. However, we confine ourselves to three further observations.

We ask how many ways an integer n may be written in the form $a^2 + b^2 + c^2, a \geq b \geq c \geq 0$. Letting S be the set of n having exactly one such representation it is clear that $n \in S$ if and only if $4n \in S$ and also that for $n \equiv 7 \pmod 8$, n is not in S.

Bateman and Grosswald proved the following two results.

Theorem 7.1. *If $n \equiv 3$ (mod 8), then $n \in S$ if and only if n is one of the numbers* 3, 11, 19, 35, 43, 67, 91, 115, 163, 235, 403, 427.

Theorem 7.2. *If n is congruent to one of* $1, 2, 5, 6$ *modulo* 8 *and $n \in S$, then either*

a. *n is one of* 1, 2, 5, 6, 10, 13, 14, 21, 22, 30, 37, 42, 46, 58, 70, 78, 93,

133, 142, 190, 253 *or,*
b. $n > 10^6$, *n is square-free, and the class number of the field* $Q(\sqrt{-4n})$ *is* 4.

An integer is a sum of three squares if and only if it is not of the form $4^a(8k+7)$. For a proof of this theorem and other information on sums of squares see the book by Grosswald cited below.

Granville and Zhu proved the following theorem.

Theorem 7.3. *For every positive integer n, except* 1, 2, 3, 4, 5, 9, 14, 17, 18, 20, 21, 35, *and* 41, *there exists an integer m,* $0 \le m \le n$, *divisible by* 4, *for which* $\binom{n}{m}$ *cannot be represented as a sum of three squares.*

The following isolated result on sums of squares is due to Bromhead . There are polynomials a, b, c such that a^2+b^2, a^2+c, b^2+c, a^2+b^2+c are all squares. Taking the variable to be unity yields the numerical solution $a = 124$, $b = 957$, $c = 13852800$.

REFERENCES

[1] O. Fraser and B. Gordon, On representing a square as the sum of three squares, *Amer. Math. Monthly* 76 (1969) 922–23.

[2] P. T. Bateman and E. Grosswald, Positive integers expressible as a sum of three squares in essentially only one way, *J. Number Theory.* 19 (1984) 301–308.

[3] E. Grosswald, *Representations of Integers as Sums of Squares*, Springer, New York, 1985.

[4] A. Granville and Y. Zhu, Representing binomial coefficients as sums of squares, *Amer. Math. Monthly* 97 (1990) 486–493.

[5] T. Bromhead, On square sums of squares, *Math. Gaz.* 44 (1960) 219–220.

8 Primes as Range of a Polynomial

There exists a 5th degree polynomial, in many variables, whose positive range is precisely the set of prime numbers. By allowing the degree to increase, Matijasevič showed, in 1977, that 10 variables would suffice for such a polynomial. This result, widely thought for a long time to be fantastically wide of any possible truth, is one of the consequences of the Matijasevič proof that there is no algorithmic procedure determining whether or not an arbitrary integral polynomial in many variables does or does not have a solution. Hilbert requested, as the tenth among his 23 problems, such a procedure. (See Integer 23, Hilbert's List.)

Work by Martin Davis, Hilary Putnam, and Julia Robinson during the 1950s and early 1960s led to an important breakthrough in 1961 when they published a proof that for exponential Diophantine equations there is no decision procedure. In other words, for these equations no finite procedure can exist enabling one to determine whether or not an arbitrary one of them has a solution.

Some of the consequences of such an approach leading to a negative solution of Hilbert's tenth problem were so surprising to many that in a review of the above mentioned paper George Kreisel had the following to say.

> These results are superficially related to Hilbert's Tenth Problem on (ordinary, i.e., non-exponential) Diophantine equations. The proof of the authors' results, though very elegant, does not use recondite facts in the theory of numbers nor in the theory of r.e. sets, and so it is likely that the present result is not closely connected with Hilbert's Tenth Problem. Also it is not altogether plausible that all (ordinary) Diophantine problems are uniformly reducible to those in a fixed number of variables of fixed degrees, which would be the case if all r.e. sets were Diophantine.

Matijasevič's 1970 solution of Hilbert's tenth problem, defended as a part of his doctoral dissertation on February 24, 1972, depended in a crucial way on the result of Davis, Putnam, and Robinson and the "not altogether plausible" is now known to be true.

The crucial result of Davis, Putnam, and Robinson was, and we quote Matijasevič, "... if even one diophantine predicate had exponential growth, then every enumerable predicate would be diophantine." He went on to say, "The predicate 'v is the $2u$th Fibonacci number' has exponential growth. We will show that it is diophantine."

See also Integer 23, Hilbert's List.

REFERENCES

[1] Ju. V. Matijasevič, Primes are enumerated by a polynomial in 10 variables, *Mat. Inst. Steklov* 68 (1977) 144–45.

[2] K. Ruohonen, Hilbert's tenth problem, *Arkhimedes* (1972) 71-100.

[3] M. Davis, Hilbert's tenth problem is unsolvable, *Amer. Math. Monthly* 80 (1973) 233–269.

[4] M. Davis, Y. Matijasevič, and J. Robinson, Hilbert's tenth problem. Diophantine equations: positive aspects of a negative solution in *Proc. Symp. in Pure Math.* xxviii, Amer. Math. Soc., Providence, Pt. 2, 1976, pp. 223–278.

[5] G. Kreisel, *Math. Rev.* (1962) A3061.

[6] Ju. V. Matijasevič, Enumerable sets are diophantine, *Soviet Math. Dokl.* 11 (1970) 354–58.

9 A Recursive Sequence

Let k be any non-square integer and let x_0 be an arbitrary rational number. For $n > 0$ define

$$x_n = \frac{x_{n-1} + 1}{\frac{x_{n-1}}{k} + 1}.$$

Stolarsky showed that for $k = 2, 3$ infinitely many of the x_n are convergent to the simple continued fraction expansion of $\sqrt{k}$. Hensley showed that the same is true for $k = 5$ but for no integer greater than 5. Hensley observed the result has an application in relativity theory.

REFERENCES

D. Hensley, Simple continued fractions and special relativity, *Proc. Amer. Math. Soc.* 67 (1977) 219–220.

10 Numbers of the Form $n! + k$

$n! + k$ is a prime or a power of a prime for exactly five pairs (n, k), $n \geq 2$, $2 \leq k \leq n$. These five pairs are $(2,2)$, $(3,2)$, $(3,3)$, $(4,3)$, $(5,5)$.

When $k = 1$ the quantity $n! + 1$, being divisible by no prime less than n, shows us that, since every integer is a product of primes, there is a prime larger than n. Thus we have a proof of the infinitude of primes.

It is not known how many of the numbers $n! + 1$ are actually prime themselves. Buhler, Crandall, and Penk have shown that the number is prime for

$$n = 1,\ 2,\ 3,\ 11,\ 27,\ 37,\ 41,\ 73,\ 77,\ 116,\ 154,\ 320,\ 340,\ 399,\ 427$$

and composite for all other n less than 546. (In an article devoted to large primes, Yates tells us we may extend the above list of n to include 872 and 1477.)

Infinitely many of these numbers with odd n are composite. Moreover, Schinzel has shown that $cn! + 1$, c an arbitrary rational number, is composite for infinitely many n.

In Euclid's proof of the infinitude of the prime numbers, he used the product of the first n prime numbers, denoted here by P_n. He observed that $E_n = P_n + 1$ is divisible by no prime smaller than the nth prime and, therefore, there are more than n primes. It is not known whether this quantity E_n is infinitely often prime. It is also not known if this quantity is infinitely often composite.

According to Buhler, Crandall, and Penk, E_n is prime for

$$p_n = 2,\ 3,\ 5,\ 7,\ 11,\ 31,\ 379,\ 1019,\ 1021,\ 2657$$

and is composite for all other p_n less than 3088. (Yates added 3229, 4547, and 4787 to this list and says there are no more up to 10133.) They also discussed primality of $n! - 1$ and $P_n - 1$ and gave similar results.

Erdős and Stewart have shown that the equations

$$P_n = x^n \pm y^m$$

are not realizable with $x, y, n > 2$, $m > 1$. (For $n = 2$, $P_2 = 1^m + 1^m$, for all m.)

Golomb reports the following.

> The anthropologist Reo Fortune (once married to Margaret Mead) conjectured that if Q_n is the smallest prime number strictly greater than E_n, then the difference $Q_n - P_n$ is *always* prime.

According to Golomb, this conjecture first appeared in print in Martin Gardner's *Scientific American* column of December 1980. He believes it "quite likely" to be true since each of the following two conjectures implies its truth.

Conjecture 1 (Cramér). *The gap between consecutive primes near x never exceeds* $(\log x)^2$.

Conjecture 2. *There is always a prime between* $P_n + 1$ *and* $P_n + p_{n+1}^2$.

There is a result of Huxley and Iwaniec reminiscent of this last and of considerable interest in its own right. It is: there is a $\theta < 1$ such that for all x sufficiently large the interval $[x, x + x^\theta]$ contains a prime number of the form $m^2 + n^2 + 1$.

In fact, Huxley and Iwaniec gave the order of the number of such primes as

$$\frac{x^\theta}{(\log x)^{\frac{3}{2}}}.$$

They offered $\frac{99}{100}$ as a permissible value of θ. This value has been lowered since 1975.

See also Integer 654, Gaps in Primes, and Integers 714, 715.

REFERENCES

[1] T. Grundhöfer, Über die Zahlender Form $n! + k$, *Arch. Math.* 33 (1972) 131–34.

[2] J. P. Buhler, R. E. Crandall and M. A. Penk, Primes of the form $n! \pm 1$ and $2 \cdot 3 \cdot 5 \cdots p \pm 1$, *Math. Comp.* 38 (1982) 639–643.

[3] S. Yates, Sinkers of the titanics, *J. Rec. Math.* 17 (1984/85) 268–274.

[4] A. Schinzel, On the composite integers of the form $c(ak+b)!\pm 1$, *Nord. Mat. Tids.* 10 (1962) 8–10.

[5] P. Erdős and C. L. Stewart , On the greatest and least prime factors of $n! + 1$, *J. London Math. Soc.* 13 (1976) 513–519.

[6] S. Golomb, The evidence for Fortune's conjecture, *Math. Mag.* 54 (1981) 209–210.

[7] M. N. Huxley and H. Iwaniec , Bombieri's theorem in short intervals, *Mathematika* 22 (1975) 188–194.

II Duffinian Numbers

In "The Duffinian Numbers," L. R. Duffy said the following.

> It often turns out that much fun can be had by defining new classes of numbers or other mathematical objects and investigating their resulting mathematical properties and relations. This article describes such a venture into the elementary theory of numbers.

A positive composite integer is *Duffinian* if the sum of its factors (other than itself) is not divisible by any of those factors (except 1).

Heichelheim showed that there exist 5 consecutive Duffinian numbers but there do not exist longer stretches of consecutive Duffinian numbers. The least such string is A, $A+1$, $A+2$, $A+3$, $A+4$, where $A = 202605639573839041$.

REFERENCES

[1] L. R. Duffy, The Duffinian numbers, *J. Rec. Math.* 12 (1979/80) 112–115.

[2] P. Heichelheim, There exist five consecutive Duffinian integers but not six, *J. Rec. Math.* 14 (1981/82) 25–28.

12 Digital Roots

If one successively adds the base 10 digits of an integer n until one arrives at a single digit, this single digit is called the *digital root* of n and is denoted by $\rho(n)$. For example $\rho(745) = 7$. Some observations about this function yield such equalities as

$$\rho(x + 9) = \rho(x), \qquad \rho(a \pm b) = \rho(\rho(a) \pm \rho(b)),$$
$$\rho(ab) = \rho(\rho(a)\rho(b)).$$

For each given n_1, define n_k, $k \geq 2$, recursively by:

$$n_k = n_{k-1} + \rho(n_{k-1}).$$

Thus $n_k = n_1 + \sum_{j=1}^{k-1} \rho(n_j)$. For $n_1 = 1$ the sequence $\{n_j\}$ is:

$$1, 2, 4, 8, 16, 23, 28, 29, 31, 35, 43, 50, 55,$$
$$56, 58, 62, 70, 77, 82, 83, 85, 89, 97, \ldots$$

and the corresponding digital roots are

$$1, 2, 4, 8, 7, 5, 1, 2, 4, 8, 7, 5, \ldots.$$

For n_1 equal to 3, 5, 7, 9, respectively, the sequences of digital roots are:

$$3, 6, 3, 6, 3, \ldots$$
$$5, 1, 2, 4, 8, 7, 5, 1, 2, 4, 8, 7, \ldots$$
$$7, 5, 1, 2, 4, 8, 7, 5, 1, 2, 4, 8, \ldots$$
$$9, 9, 9, \ldots.$$

Kumar showed that no matter what we take our starting integer n_1 to be the corresponding sequence of digital roots will always end with one of the 5 displayed above.

Asadulla noted that the digital root of the nth Fermat number $2^{2^n} + 1$ is 5 or 8 according to whether n is odd or even.

Brooke argued that with the exception of 6 all perfect numbers have digital root equal to 1.

REFERENCES

[1] V. S. Kumar, On the digital root series, *J. Rec. Math.* 12 (1979/80) 267–270.

[2] S. Asadulla, A note on Fermat numbers, *J. Natur. Sci. and Math.* 17 (1977) 113–118.

[3] M. Brooke, On the digital roots of perfect numbers, *Math. Mag.* 34 (1960/61) 100, 124.

13 The Banach–Tarski Paradox

The Banach–Tarski paradox (actually *theorem*), first proved in 1924, asserts that the unit sphere may be decomposed into a finite number of pieces which may then be reassembled into two unit spheres.

In 1929 von Neumann asserted, without proof, that the decomposition could be effected using 9 pieces, and, in 1945, Sierpiński proved that 8 pieces suffice. Then, in 1957, Robinson proved that this could be done using exactly 5 pieces, one a single point, and, further, that no fewer than 5 pieces would suffice even if reflections are used.

Thus it is possible to decompose a sphere S of radius 1 into 5 pieces S_1, S_2, S_3, S_4, S_5 so that for suitable rigid motions R_1, R_2, R_3, R_4, R_5 the three sets

$$S; \quad R_1(S_1) \cup R_2(S_2); \quad R_3(S_3) \cup R_4(S_4) \cup R_5(S_5)$$

are congruent.

An alternate formulation of the theorem is: *Given $\epsilon > 0$, $M > 0$, and two spheres S, S' of radii ϵ, M, respectively, there is a decomposition of S' into a finite number of pieces $S_1, \ldots, S_n$ such that for suitable rigid motions $R_1, \ldots, R_n$ it is true that*

$$S = R_1(S_1) \cup \cdots \cup R_n(S_n).$$

Thus, as E. Kasner and J. Newman once observed, one can decompose the sum into a finite number of pieces, reassemble them into a marble-sized sphere, and then pocket the sphere.

Quite recently Laczkovich solved a problem first posed by Tarski in 1925. As stated by Gardner and Wagon, the problem reads as follows.

> Is it possible to partition a circle (with interior) into finitely many sets that can be rearranged (using isometries) so that they form a partition of a square? In short: Is a circle piece-wise congruent to a square?

The paper by Gardner and Wagon presents the history of the problem and its connection with the problem of equidecomposability discussed in Integers 2, 3, Equidecomposability.

Two observations in the paper are:

1. "... the number of pieces needed to square an isosceles right triangle is about 10^{50}.";
2. "Finally, even though it is too late to gain immortality by squaring the circle or trisecting the angle, you can try trisecting the circle. The problem is to partition a disk into three congruent sets. An extension of the Banach–Tarski Paradox discovered by Raphael Robinson ... shows that a ball may be trisected."

See also Integers 2, 3, Sets Congruent to a Proper Part.

REFERENCES

[1] W. Sierpiński, *On the Congruence of Sets and their Equivalence by Finite Decomposition*, Lucknow Univ. Studies xx, Lucknow, 1954.

[2] R. M. Robinson, On the decomposition of spheres, *Fund. Math.* 34 (1947) 246–260.

[3] S. Wagon, *The Banach–Tarski Paradox*, Cambridge Univ. Press, Cambridge, 1985.

[4] E. Kasner and J. Newman, *Mathematics and the Imagination,* Simon and Schuster, New York, 1940, p. 207.

[5] R. J. Gardner and S. Wagon, At long last, the circle has been squared, *Notices of the AMS* 36 (1989) 1338–1343.

14 Euclidean Algorithm

In carrying out the Euclidean algorithm to find the greatest common divisor of two positive integers a, b, the number of steps needed will never exceed 5 times the number of base 10 digits in the smaller of the two integers a, b.

Various aspects of this theorem, first proved by Lamé in 1844, are quite regularly rediscovered. No doubt a "natural high" occurs each time this happens. (At least that was so in my case.)

REFERENCES

[1] J. V. Uspensky and M. A. Heaslet , *Elementary Number Theory*, McGraw-Hill, New York, 1939, pp. 43–45.

15 Sieve Approaches to Twin Primes

Using his sieve, Selberg has shown the existence of infinitely many n for which $n(n + 2)$ is the product of at most 5 primes.

Porter has given the following similar results.

1. There are infinitely many n such that $(8n + 1)(n^2 + n + 1)$ is the product of at most 6 prime factors.
2. There are infinitely many primes p such that $(p + 2)(p + 6)$ is the product of at most 7 prime factors.
3. There are infinitely many primes p such that $(p + 2)(p^2 + p + 1)$ is the product of at most 9 prime factors.

In a (not quite) similar vein, Vaughan showed the following.

Either: There exist infinitely many primes p such that $8p + 1$ is the product of at most 2 primes.

Or: There exist infinitely many n such that n and $n + 1$ have the same number, $\tau(n)$, of divisors. (Erdős and Mirsky had asked if this was true in 1952.)

In a 1981 thesis, Spiro proved the existence of infinitely many n for which $\tau(n) = \tau(n + 5040)$. Motivated by her work, Heath-Brown proved, in 1984, that there are infinitely many integers for which $\tau(n) = \tau(n+1)$. In fact, he proved that if $N(x)$ is the number of n not exceeding x for which the equality holds, then

$$N(x) >> \frac{x}{(\log x)^7}.$$

(Later, a master's student of his, C. Pinner, showed the 1 can be replaced by an arbitrary positive integer.) Three years later, Hildebrand improved this result by replacing the 7 by 3.

In the paper by Heath-Brown, his *Key Lemma* is of some independent interest. It reads as follows.

For every positive integer M there exist M integers a_i, $a_1 < a_2 < \cdots < a_M$, such that if we put $a_{mn} = a_m - a_n$ for $m > n$, then a_{mn} divides each of a_m and a_n and, furthermore,

$$\tau(a_m)\tau\left(\frac{a_n}{a_{mn}}\right) = \tau(a_n)\tau\left(\frac{a_m}{a_{mn}}\right).$$

The following assertion is an easy exercise. Each of the two statements,

$$\sigma(n) = n + 1 + 2\sqrt{n+1} \qquad \text{and} \qquad \phi(n) = n + 1 - 2\sqrt{n+1},$$

is equivalent to the assertion that n is a product of twin primes.

REFERENCES

[1] J. W. Porter, Some numerical results in the Selberg sieve method, *Acta Arith.* 20 (1972) 417–421.

[2] R. C. Vaughan, A remark on the divisor function $d(n)$, *Glasgow Math. J.* 14 (1973) 54–55.

[3] S. A. Sergušov, On the problem of twin primes (in Russian), *Jaroslav. Gos. Ped. Inst. Učen. Zap. Vyp.* 82 (1971) 85–86.

[4] C. A. Spiro, The frequency with which an integral-valued, prime-independent, multiplicative or additive function of n divides a polynomial function of n, PhD thesis, U. of Ill., 1981.

[5] D. R. Heath-Brown, The divisor function at consecutive integers, *Mathematika* 31 (1984) 141–49.

[6] A. Hildebrand, The divisor function at consecutive integers, *Pac. J. Math.* 129 (1987) 307–319.

16 An Extremal Property of 5

Let $S_q(N)$ be the sum of all of the digits appearing in the base q representations of the numbers $1, \ldots, N$ and let $K_q(N)$ be the number of digits in the base q representation of N. Further, put

$$f(q, N) = \frac{(K_q(N) + 1)S_q(N)}{N}.$$

Then, for N sufficiently large,

$$f(5, N) < f(q, N)$$

whenever $q \neq 5$.

REFERENCES

G. Hofmeister and H. Waadeland , Eine Minimaleigenschaften des Fünfer-Systems, *Norske. Vid. Selsk. Forh.* 39 (1966) 66–72.

17 Volumes of n-Dimensional Spheres

Let $u_n(a)$ denote the volume of the n-dimensional sphere of radius a. Then

$$u_n(a) = \int \cdots \int_{x_1^2 + \cdots + x_n^2 < a^2} dx_1 \cdots dx_n.$$

It is not hard to show that $u_n(a) = a^n u_n(1)$ and that

$$
\begin{aligned}
u_n(1) &= \int\!\!\int_{x_1^2+x_2^2<1} \left\{ \int \cdots \int_{x_3^2+\cdots+x_n^2<1-x_1^2-x_2^2} dx_3 \cdots dx_n \right\} dx_1\, dx_2 \\
&= u_{n-2}(1) \int\!\!\int_{x_1^2+x_2^2<1} \left(1 - x_1^2 - x_2^2\right)^{(n/2)-1} dx_1\, dx_2 \\
&= \frac{2\pi}{n} u_{n-2}(1).
\end{aligned}
$$

Simple calculations yield $u_2(1) = \pi$, $u_3(1) = \frac{4\pi}{3}$. From these the recurrence gives rise to

$$
u_n(1) = \begin{cases} \dfrac{(2\pi)^{n/2}}{n(n-2)(n-4)\cdots 2}, & \text{for } n \text{ even} \\ \dfrac{2(2\pi)^{(n-1)/2}}{n(n-2)(n-4)\cdots 1}, & \text{for } n \text{ odd.} \end{cases}
$$

Using the gamma function this becomes

$$
u_n(1) = \frac{\pi^{n/2}}{\Gamma(\frac{n}{2} + 1)}.
$$

Utilizing this we see that as n tends to infinity $u_n(1)$ tends to zero. Further, we see that $u_n(1)$ increases as n goes from 1 to 5 and then decreases. Thus

> the unit sphere in 5 dimensions has larger volume than the unit sphere in any other dimension.

The value of $u_5(1)$, the volume of the unit sphere in 5 dimensions, is $\frac{8\pi^2}{15}$.

A recent, accessible discussion of this phenomenon is given by Smith and Vamanamurthy. In this connection, see also Sagan.

REFERENCES

[1] D. J. Smith and M. K. Vamanamurthy, How small is a unit ball?, *Math. Mag.* 62 (1989) 101–107.

[2] H. Sagan, Letter to the Editor, *Math. Mag.* 63 (1990) 205.

18 Regular Polyhedra

A *regular polyhedron* is a convex polyhedron with congruent regular polygons having, say, p (≥ 3) edges as faces and having at each vertex the same number, say q (≥ 3) edges. We say such a polyhedron is of type $\{p, q\}$.

Let f, e, v be the respective numbers of faces, edges, and vertices of a regular polyhedron of type $\{p, q\}$. Then, using Euler's relation $v - e + f = 2$, we find

$$\frac{1}{p} + \frac{1}{q} = \frac{1}{2} + \frac{1}{e}.$$

Now $p \geq 6$ would imply $q < 3$, contrary to fact, so $p = 3$, 4, or 5. Similarly these are the candidates for q. Of the 9 combinations only 5 are possible since for the others the sums of the reciprocals of p and q are not greater than a half.

The above argument only gives an upper bound on the number of possible regular polyhedra. It is not difficult to convince ourselves that in fact each of the five types is realizable. For an interesting discussion with further references, see the source cited below.

The five types are: $\{3, 3\}$ tetrahedron, $\{3, 4\}$ octahedron, $\{3, 5\}$ icosahedron, $\{4, 3\}$ cube, and $\{5, 3\}$ dodecahedron.

REFERENCES

A. Beck, M. N. Bleicher, and D. W. Crowe, *Excursions into Mathematics*, Worth Pub., New York, 1972, Chapter 1.

Integer 6

1 Sum of Distinct Primes

Richert showed the integer 6 is the largest integer which is neither a prime nor the sum of two or more distinct primes.

All integers other than 1, 2, 4, 6, 9 may be written as the sum of distinct odd primes.

Every integer larger than 45 is a sum of distinct primes each of which is greater than or equal to 11.

Every integer greater than 57 is a sum of distinct primes each of which is greater than or equal to 13.

A generalization of Richert's theorem given by Porubský is as follows. Let $f(x)$ be an integral polynomial such that for every integer $d > 1$ there are infinitely many primes p with $f(p)$ not divisible by d. Then every sufficiently large integer is representable as a sum of distinct elements of the set $\{f(p)|p \text{ is prime}\}$.

Kløve let $a(q)$ stand for the largest integer not a sum of distinct primes each greater or equal to q. (By Richert, $a(2) = 6$.) He also gave a table of values of this function, a small part of which is produced below.

q	2	3	5	7	11	13	17	19	23	29	7927
$a(q)$	6	9	27	45	45	57	75	81	87	105	23867

Kløve conjectured that $a(q)$ is asymptotic to $3q$ and that $a(q)$ is odd for $q > 2$. See also Integer 9, Sums of Distinct Odd Primes and Integer 33, Complete Sequences.

REFERENCES

[1] H. E. Richert, Über zerfällungenin ungleiche Primzahlen, *Math. Zeit.* 52 (1941) 342–43.

[2] R. E. Dressler, A stronger Bertrand's postulate with an application to partitions, *Proc. Amer. Math. Soc.* 33 (1972) 226–28.

[3] R. E. Dressler, Sums of distinct primes, *Nord. Mat. Tidsskr.* 21 (1973) 31–32.

[4] J. L. Brown, Generalization of Richert's theorem, *Amer. Math. Monthly* 83 (1976) 631–634.

[5] Š. Porubský, Sums of prime powers, *Monatsh. Math.* 86 (1978/79) 301–303.

[6] T. Kløve, Sums of distinct primes, *Nord. Math. Tidskr.* 21 (1973) 138–140.

2 Values of ϕ

Vaidya has shown there are exactly two values of n for which $\phi(n) \not\geq \sqrt{n}$ and they are 2 and 6. For all n, $\phi(n) \geq \sqrt{\frac{n}{2}}$.

Denote the number of n for which $\phi(n) = m$ by $a(m)$ and put $A(x) = \sum_{m \leq x} a(m)$. Though the function $a(m)$ is erratic $A(x)$ is quite regular for large x. Subbarao and Yip observed, for instance, that $a(1438) = 2$, $a(1442) = 72$, $a(1440) = a(1444) = 0$. Nevertheless, in 1945, Erdös showed that $\lim \frac{A(x)}{x}$, as $x \to \infty$, exists, and, in 1970, Dressler showed the limit to be

$$\frac{\zeta(2)\zeta(3)}{\zeta(6)}.$$

In 1972 Bateman gave simple proofs of these facts.

Montgomery uses the following formula.

$$\sum_{n \leq x} \frac{1}{\phi(n)} = \frac{\zeta(2)\zeta(3)}{\zeta(6)} \log x + A + O(\frac{\log x}{x})$$

The above expression involving the zeta function also occurs in an *American Mathematical Monthly* problem posed by Erdős and Motzkin and solved by Bumby. If $F(n)$ is the number of pairs of positive integers $a,\ b$, that do not exceed n, that have the same prime factors, then

$$\frac{F(n)}{n} \rightarrow \frac{\zeta(2)\zeta(3)}{\zeta(6)}.$$

The sequence $\{\frac{\phi(n)}{n}\}$ is dense in the unit interval. Generalizations of this have been given by Porubský [7].

Subbarao and Yip state that if m is even and all three of the numbers $\frac{1}{2}m,\ m+1,\ 2m+3$ are prime, then $\phi(\phi(x)) = m$ has exactly two solutions and they are $x = 2m+3,\ 4m+6$. In the case where $m = 10$, we see that $x = 23$ or $x = 46$. In this case we see that when $\phi(y) = 10$, $y = 11$ or $y = 22$.

Klee showed that the equation $\phi(x) = 2m$, where m is odd and larger than 1, has either 0, 2, or 4 solutions.

REFERENCES

[1] A. M. Vaidya, An inequality for Euler's totient function, *Math. Stu.* 35 (1967) 79–80.

[2] M. V. Subbarao and L. W. Yip, Carmichael's conjecture and some analogues in *Number Theory* (eds. J.-M. Dekoninck and C. Levesque), de Gruyter, New York, 1989, pp. 928–941.

[3] P. Erdős, Some remarks on Euler's ϕ function and some related problems, *Bull. Amer. Math. Soc.* 51 (1945) 540–544.

[4] P. T. Bateman, The distribution of values of the Euler function, *Acta Arith.* 21 (1972) 329-345.

[5] H. L. Montgomery, Primes in arithmetic progression, *Mich. Math. J.* 17 (1970) 33-39.

[6] R. T. Bumby, Adv. Prob. 5735, *Amer. Math. Monthly* 78 (1971) 680–81. (Proposed by P. Erdős and T. Motzkin.)

[7] Š. Porubský, Über die Dichtigkeit der Werte multiplikativer Funktionen, *Math. Slovaca* 29 (1979) 69–72.

[8] V. L. Klee, On the equation $\phi(x) = 2m$, *Amer. Math. Monthly* 53 (1946) 327–28.

3 Divisors of $a^n - b^n$

Let a, b, n be positive integers with $a > b$. We say a prime p is *primitive* for $a^n - b^n$ if p integrally divides this quantity, but does not integrally divide $a^m - b^m$ for any positive m smaller than n.

With the exception of $n = 6$, $a^n - b^n$ has, for every $n > 2$, a prime primitive. This was proved by Zsigmondy in 1892 and by Bang in 1896. For $n = 6$, we have $2^6 - 1^6 = 63 = 3^2 \cdot 7$. Here 3 divides $2^2 - 1^2$ and 7 divides $2^3 - 1^3$ so every prime divisor of $2^6 - 1^6$ divides $2^m - 1^m$ for some m, $0 < m < 6$.

If we let $P(m)$ be the largest prime divisor of m, then there is a set of integers of density 1 such that the quotient

$$\frac{P(a^n - b^n)}{n}$$

tends to infinity as n runs over the set. In fact a little more is the case. When $a \neq b$ and p is prime, $P(a^p - b^p) > Cp \log p$ for some constant C independent of p.

In 1985 Sun Qi and Zhang Mingzhi showed that if $0 \leq b \leq a$, then $2^a - 2^b$ divides $n^a - n^b$ for all n if and only if (a, b) is one of the 14 pairs $(1, 0)$, $(2, 1)$, $(3, 1)$, $(4, 2)$, $(5, 3)$, $(5, 1)$, $(6, 2)$, $(7, 3)$, $(8, 4)$, $(8, 2)$, $(9, 3)$, $(14, 2)$, $(15, 3)$, $(16, 4)$.

REFERENCES

[1] G. D. Birkhoff and H. S. Vandiver, On the integral divisors of $a^n - b^n$, *Ann. Math.* 5 (1903/4) 173–180.

[2] L. P. Postnikova and A. Schinzel, Primitive divisors of the expression $a^n - b^n$ in algebraic number fields (Russian), *Mat. Sb. (N.S.)* 75 (117) (1968) 171–77.

[3] C. L. Stewart, On divisors of Fermat, Fibonacci, Lucas, and Lehmer numbers, *Proc. London Math. Soc.* 35 (1977) 425–447.

[4] T. N. Shorey and C. L. Stewart, On divisors of Fermat, Fibonacci, Lucas and Lehmer numbers II, *J. London Math. Soc.* 23 (1981) 17–23.

[5] S. Qi and Z. Mingzhi, Pairs where $2^a - 2^b$ divides $n^a - n^b$ for all n, *Proc. Amer. Math. Soc.* 93 (1985) 218–220.

4 Square Permutations

A permutation P of a set of integers is called a *square permutation* if $k + P(k)$ is a square for all k in the set of integers.

If

$$P = (1\ 3\ 6\ 10\ 15)(2\ 7)(4\ 5)(8\ 17)(9\ 16)(11\ 14)(12\ 13),$$

then P is a square permutation on the set of the first 17 positive integers as may be seen readily by examining consecutive terms in the individual cycles of the permutation.

If

$$P = (0\ 4)(1\ 8)(2\ 14)(3\ 6)(5\ 11)(7\ 9)(10\ 15)(12\ 13),$$

we again see that P is a square permutation. This time the permutation is on the first 16 *nonnegative* integers.

It is an interesting fact that square permutations exist for permutations on the first n nonnegative integers for all nonnegative integers n, but there are exactly 6 positive integer values for which no such permutations exist on the set of the first n positive integers. The exceptional values of n are 1, 2, 4, 6, 7, 11.

In a similar fashion, one may call a permutation *prime* if $k + P(k)$ is prime for all k. Again, one can ask for the existence of prime permutations for various permutations.

REFERENCES

B. L. Schwartz, Square permutations, *Math. Mag.* 51 (1978) 64–66

5 Shrinking Labelled Squares

Consider the following operation on a square whose vertices are labelled with real numbers. At the midpoint of each side place the number which is the absolute value of the difference between the numbers which are at the ends of the side. These four midpoints with their attendant labels form a smaller labelled square. The process can now be repeated as many times as we please.

If one uses integer labels a rather surprising thing happens. One finds that, after a finite number of repetitions, one will ultimately come to a situation where all labels are 0. Even when the labels are not integers, the following are true.

- If the largest and the smallest of the labels are not adjacent, then 6 repetitions always suffice to arrive at all zero labels.
- If two nonconsecutive labels are equal, then 4 repetitions suffice.
- If no three of the labels sum to the fourth, then 7 repetitions suffice.
- There exist labellings which will, after a finite number of repetitions, produce all zero labels, but which require at least as many repetitions as one might specify.

One has to be somewhat careful since not all that is claimed is true. For instance, the following is claimed: If the 7th repetition does not result in all zero labels, then the labelling was not monotonically increasing in either the clockwise or counterclockwise sense. But the labelling $1, 3, 9, 19$ requires 8 steps while $5, 9, 17, 31$ requires 10 steps.

If one denotes a labelled square by $[a, b, c, d]$, where the a, b, c, d are the labels in, say, clockwise order, and defines the mapping T by

$$[a, b, c, d]\,T = [|a-b|, |b-c|, |c-d|, |d-a|]\,,$$

we see that this T describes our process of passing from one labelled square to the next. Using this notation and letting $v = [1, q, q^2, q^3]$, where q is the real zero of the polynomial $q^3 - q^2 - q - 1$, it is easy to see that for k any positive real number we have $(kv)T = k(q-1)v$. Consequently, no number of repetitions of the map T will take such labelled squares as kv into a square with all labels zero. However, except for these labellings every labelled square will, after a finite number of steps, lead to the square with all zero labels.

Freedman, in a 1948 paper, reported having been given (in 1938, by Jekuthiel Ginsburg) the integer version of the problem as an exercise. Apologizing for the 10 year delay, he generalized the problem as follows. Let

$$S_r = \{(a_0, \dots, a_{r-1}) | a_i \in Z, a_i \geq 0\},$$

and define $D : S_r \to S_r$ so that

$$D(a_0, \dots, a_{r-1}) = (|a_0 - a_1|, |a_1 - a_2|, \dots, |a_{r-1} - a_0|).$$

Then for $s \in S_r$ there is an n with the nth iterate of D, applied to s, consisting of all 0's if and only if n is a power of 2.

Eltermann and Svengrowski also considered this generalized problem in their papers cited below.

One might well conjecture, just by reading the titles of the papers listed below, that many investigators were unaware of much or all of the other work on the problem.

REFERENCES

[1] E. R. Berlekamp, The design of slowly shrinking labelled squares, *Math. Comp.* 29 (1975) 25–27.

[2] Z. Magyar, A recursion on quadruples, *Amer. Math. Monthly* 91 (1984) 360–362.

[3] B. Freedman, The four number game, *Scripta Math.* 14 (1948) 35–47.

[4] M. Lotan, A problem in difference sets, *Amer. Math. Monthly* 56 (1949) 535–541.

[5] P. Svengrowski, Iterated absolute values, *Math. Mag.* 52 (1979) 36–40.

[6] A. L. Furno, Cycles of differences of integers, *J. Number Theory* 13 (1981) 255–261.

[7] M. Dumont and J. Meeus, The four-numbers game, *J. Rec. Math.* 13 (1980/81) 89–96.

[8] H. Eltermann, Iterierte absolute Differenzen mit natürlichen Zahlen, *J. für Reine und Angew. Math.* 201 (1959) 71–77.

[9] R. Greenwell, The game of Diffy, *Math. Gaz.* 73 (1989) 222–225.

[10] A. Ludington-Young, Length of the n-number game, *Fib. Q.* 28 (1990) 259-265.

6 Irreducible Polynomials

Let f be an integral polynomial of degree n. If $n > 6$ and if there is a prime p such that among the values of f at least n of them are either p or $-p$, then the polynomial is either irreducible or it factors into the product of two irreducible polynomials of equal degree.

When $n = 6$ this is not the case as the following example illustrates.

$$(x+2)(x+1)x(x-1)(x-2)(x-3)-5 = (x^2-x-1)(x^4-2x^3-6x^2+7x+5)$$

REFERENCES

H. L. Dorwart, Can this polynomial be factored?, *Two Year Coll. Math. J.* 8 (1977) 67–72.

7 Bilinear Transformations

We consider the transformation

$$T(x) = \frac{ax + b}{cx + d},$$

where the a, b, c, d are integers, $ad - bc \neq 0$, and $c \neq 0$. Given a value, say x_0, we can use T to define a sequence by putting $x_{n+1} = T(x_n)$. Two cases in which the x_i are all integers are afforded by:

1. $x_1 = 2, x_{n+1} = \dfrac{x_n}{x_n - 1}$, where the sequence is 2, 2, 2, ...

2. $x_1 = 4, x_{n+1} = \dfrac{178x_n - 1492}{7x_n + 2}$, where the sequence is 4, −26, 34, 19, 14, 10, 4, ...

Independently of the other parameters, if all the x_i are integers, then the sequence $\{x_i\}$ is periodic and always has period length one of 1, 2, 3, 4, or 6.

REFERENCES

D. M. Adelman, Note on the arithmetic of bilinear transformations, *Proc. Amer. Math. Soc.* 1 (1950) 443–48.

8 Polyhedra

In *A Survey of Geometry,* Howard Eves proves that every polyhedron may be dissected into a finite number of tetrahedra. In doing this the vertices of the tetrahedra need not all be vertices of the original polyhedron. On the other hand, if it is possible to so select the tetrahedra so that all of their vertices are vertices of the original polyhedron, then Eves calls the polyhedron a *Lennes polyhedron* .

In 1911 Lennes constructed a Lennes polyhedron with 7 vertices. In 1928 Schönhardt gave an example of such a polyhedron with 6 vertices and showed that none existed with fewer than 6 vertices. In 1948 Bagemihl showed Lennes polyhedra exist for any number of vertices greater than 6.

Thus, Lennes polyhedra of n vertices exist if and only if $n \geq 6$.

REFERENCES

H. Eves, *A Survey of Geometry*, Allyn and Bacon, Boston, 1972, pp. 211–212.

9 Perfect Numbers

A positive integer—equal to the sum of those of its divisors that are smaller than itself—is *perfect*. It is not known if there are infinitely many perfect numbers. The smallest of the perfect numbers is 6. As of 1987, there were 30 perfect numbers known and the largest of them, $2^{216090}(2^{216091} - 1)$, has 130100 digits.

It is not known if there are any perfect numbers which are odd. The best bound known to this writer is that given by Brent and Cohen to the effect that an odd perfect number can not be less than 10^{160}. (However, it was reported in the *Notices of the American Mathematical Society* 23 (1976) A-55 that a lower bound is 10^{200}.)

For n to be perfect, it is necessary and sufficient that $\sigma(n) = 2n$. It is not hard to show that for infinitely many n both inequalities, $\sigma(n) < 2n$ and $\sigma(n) > 2n$, are realizable. (See Sierpiński p. 185.)

See also Integer 28, Special Perfect Number and Integer 90, Biunitary Perfect Numbers.

REFERENCES

[1] W. Sierpiński, *Elementary Theory of Numbers,* North-Holland, Amsterdam, 1988.

[2] R. P. Brent and G. L. Cohen, A new lower bound for odd perfect numbers, *Math. Comp.* 53 (1989) 431–37.

10 Mod N Weak Uniform Distribution

For $N \geq 3$ we call the integer sequence $\{a_j\}$, having infinitely many terms relatively prime to N, *weakly uniformly distributed modulo N,* if, for all pairs of integers s, t relatively prime to N, the ratio of the number of those among the first x of the a_j relatively prime to s to the number relatively prime to t tends to 1 as x tends to infinity.

Śliva shows that the sequence $\{\sigma(n)\}$, where σ is the usual sum of divisors function, is weakly uniformly distributed (mod N) if and only if 6 does not divide N.

The definition of a weakly uniformly distributed sequence modulo N is due to Narkiewicz.

REFERENCES

[1] J. Śliva, On distribution of values of $\sigma(n)$ in residue classes, *Colloq. Math.* 27 (1973) 283–291, 332.

[2] W. Narkiewicz, On distribution of values of multiplicative functions in residue classes, *Acta Arith.* 12 (1966/67) 269–279.

II Sophie Germain's Theorem

We give the following quote from a paper by Granville.

> In 1823, Sophie Germain (*Mem. Acad. Sci. Inst. France* 6 (1823), 1–60) showed that if p, $2p+1$ are both odd primes, then the so-called "First Case" of Fermat's last theorem holds for p. This was extended by Legendre, Wendt, Vandiver, Denes , and others to prime pairs p, $mp+1$, where $6 \nmid m$ and p is sufficiently large (depending on m). The cases where 6 divides m are fraught with an inescapable technical difficulty, and, as we shall see in this paper, it requires quite sophisticated techniques to even find a partial resolution for prime pairs p, $6p + 1$.

See also Integer 1093, Fermat's Conjecture.

REFERENCES

A. Granville, Sophie Germain's theorem for prime pairs p, $6p + 1$, *J. Number Theory* 27 (1987) 63–72.

12 A Curious Sequence

For k a positive integer, consider finite sequences of k nonnegative integers

$$n_0, n_1, \ldots, n_{k-1}$$

such that each n_j is the number of occurrences of j among these k terms. (If 0 does not appear, then $n_0 = 0$, and we see that $n_0 \neq 0$. If $n_0 = k$, then all $n_j = 0$, and we see that $n_0 \neq k$. Also the sum of the k terms must equal k.) The following are true statements concerning such sequences.

1. No such sequences exist for $k = 1, 2, 3, 6$.
2. There are two such sequences for $k = 4$, and they are 1210 and 2020.
3. There is one such sequence for $k = 5$, and it is 21200.
4. There is one such sequence for $k = 7$, and it is 3211000.
5. For $k > 7$ there is exactly one such sequence, and it is

$$\{k-4\}210\ldots010000$$

where the last displayed 1 is in the $k-4$th spot.

REFERENCES

S. Kahan, A curious sequence, *Math. Mag.* 48 (1975) 290–92.

Integer 7

1 Incomparable Rectangles

Two rectangles are said to be *incomparable* if neither of them may be placed inside the other in such a way that corresponding sides are parallel.

Except for a finite number of shapes of rectangles, every rectangle may be exactly covered with 7 pairwise incomparable rectangles, but no rectangle may be exactly covered with 6 such rectangles.

Aside from rotations and reflections the pattern of a 7 piece tiling is unique.

For each $t \geq 7$, there are only finitely many integral squares that can not be tiled with t incomparable integral rectangles.

Every integral square with side greater than or equal to 34 may be tiled with 7 incomparable integral rectangles.

For all $k \geq 1$ and $N \geq 4k - 2$, an $(N + k) \times N$ integral rectangle can be tiled with 7 incomparable integral rectangles.

References

[1] A. C. C. Yao, E. M. Reingold and B. Sands, Tiling with incomparable rectangles, *J. Rec. Math.* 8 (1975/76) 112–19.

[2] R. K. Guy, A couple of cubic conundrums, *Amer. Math. Monthly* 91 (1984) 624–29.

2 Minimal Numbers

The least positive integer having exactly h divisors is denoted by $A(h)$. Letting $\tau(n)$ be the number of divisors of n we say that n is *minimal* when $A(\tau(n)) = n$; i.e., if n is the least positive integer having the number of divisors it has.

$N!$ is minimal if and only if $1 \leq N \leq 7$.

There exist exactly 14 values of N for which the least common multiple of all integers from 1 to N is minimal and these integers are 1, 2, 3, 4, 5, 6, 8, 9, 10, 11, 12, 16, 27, 28.

REFERENCES

M. E. Grost, The smallest number with a given number of divisors, *Amer. Math. Monthly* 75 (1968) 725–29.

3 A Square

"Fermat posed the following question to Wallis: Does there exist a natural number n such that $\sigma(n)$, the sum of the divisors of n, is a perfect square?" Thus starts Halberstam's review of a paper by Schinzel in which the following is proved.

The only prime p such that $\sigma(p^3)$ is a square is $p = 7$.

That is, the only prime number p for which $1 + p + p^2 + p^3$ is a square is the prime number 7. In that case we have $1 + 7 + 7^2 + 7^3 = 20^2$.

See also Integer 3, $\sigma(p^4)$.

In fact, the only positive odd numbers m for which $1 + m + m^2 + m^3$ is a square are 1 and 7. Note that it is not always the case that this sum is equal to $\sigma(m^3)$.

More generally, it appears that in 1877 Gerono showed that the equation

$$1 + x + x^2 + x^3 = y^2$$

has only the pairs $(x,y) = (-1,0),(0,\pm 1),(1,\pm 2),(7,\pm 20)$ as solutions.

Ljunggren (in 1943) extended this to show that the equation

$$1 + x + x^2 + \cdots + x^{n-1} = y^q$$

has, for $q = 2$, only the solutions $(n,x,y) = (4,7,20),(5,3,11)$ with x,y integers, $|x| > 1$. The case where q is a multiple of 3 is only possible when $q = 3$ and then only when $(n,x,y) = (3,18,7),(3,-19,7)$.

(It is interesting to note that, not knowing of Ljunggren's result, Takaku (*Colloq. Math.* 49 (1984) 117–121) proved the weaker result that if α is odd and greater than 3, then $\sigma(p^\alpha)$ being a perfect square implies p must be smaller than $2^{2^{\alpha+1}}$. Three years later, Chidambaraswami and Krishnaiah (*Proc. Amer. Math. Soc.* 101 (1987) 625–28) extended Takaku's result to almost all even integers p (still weaker than the result of Ljunggren). It is not at all unusual for this to happen in mathematics (and elsewhere?). As much as anything else it is indicative of the nonlinear character of the subject as well as the lack of universal knowledge on the part of each practitioner. It is not at all clear that such universal knowledge would make the subject more humane. Indeed, the effort at searching the literature before communicating any result, if universally insisted upon, would take a good deal of joy from the subject and, at the same time, further inflate the significance of a published paper. Thus, from our point of view, despite the fact that Ljunggren's result implies these later conclusions in a vacuous way there is little reason to regard their work as wasted. Further, Takaku later (*Colloq. Math.* 52 (1987) 319–323)) continues his investigation and extends the results beyond the range of cases covered by Ljunggren's work.)

Suryanarayana showed, in 1967, that if p is prime and q is an integer greater than 1, then $1 + q + q^2 + \cdots + q^{\mu-1}$ can be a power of p only under the conditions that $\mu = 2$, when $p = 2$, otherwise μ is a prime divisor of $p - 1$.

Edgar mentions the special equations:

1. $1 + 18 + 18^2 = 7^3$,
2. $1 + 3 + 3^2 + 3^3 + 3^4 = 11^2$.

Schinzel and Tijdeman have investigated equations of the following type:

$$y^m = P(x),$$

where $P(x)$ is a rational polynomial. They proved the following.

Theorem. *If the rational polynomial $P(x)$ has two distinct zeros, then $y^m = P(x)$, x, y integers, $|y| > 1$ implies that m does not exceed some absolute constant depending only on the polynomial $P(x)$.*

Corollary 1. *If the rational polynomial $P(x)$ has at least two simple zeros, then $y^m = P(x)$ $(x, y$ integers, $|y| > 1)$ has only finitely many integer solutions m, x, y, $m > 2$, $|y| > 1$, and they may be found effectively.*

Corollary 2. *If, as in the first corollary, there are at least three simple zeros, then the same conclusion holds with $m > 1$ replacing $m > 2$.*

They formulated the following conjecture.

Conjecture. *If the rational polynomial $P(x)$ has at least three simple zeros, then the equation $y^2 z^3 = P(x)$ has only finitely many integer solutions m, x, y, $m > 1$, $|y| > 1$ and they may be found effectively.*

The authors observed: "This conjecture lies rather deep, since it implies the existence of infinitely many primes p such that $2^{p-1} \not\equiv 1 \pmod{p^2}$." (See Integer 1093, Fermat's Conjecture.)

REFERENCES

[1] H. Halberstam, *Math. Rev.* 22 #1538 (A30-16).

[2] A. Schinzel, On prime numbers such that the sums of the divisors of their cubes are perfect squares, *Wiadom. Mat.* (2) 1 (1955/56) 203–2044.

[3] È. T. Avanesov, A certain property of the number seven (Bulgarian), *Fiz.-Mat. Spiz. B'lgar. Akad. Nauk.* 11 (44) (1968) 277.

[4] W. Ljunggren, Some theorems of the form $\frac{x^n-1}{x-1} = y^q$, *Norsk Mat. Tidsskr.* 25 (1943) 17–20.

[5] D. Suryanarayana, Certain Diophantine equations, *Math. Stu.* 35 (1967) 197–99.

[6] H. M. Edgar, The exponential Diophantine equation $1 + a + a^2 + \cdots + a^{x-1} = p^y$, *Amer. Math. Monthly* 81 (1974) 758–59.

[7] A. Schinzel and R. Tijdeman, On the equation $y^m = P(x)$, *Acta Arith.* 31 (1976) 199–204.

4 Iterates

Consider the sequence

$$1, 3, 4, 5, 6, 6, 8, 7, 7, 8, 12, \ldots$$

where the nth term, $f(n)$, is 1 plus the sum of the prime factors (each counted according to its multiplicity) of n. Then, if $n > 6$, the sequence

$$n,\ f(n),\ f(f(n)),\ f(f(f(n))), \ldots$$

is periodic with period $(7, 8)$.

This result is given in a more general form in Burkard [3, p. 249] and was unknown to Roberts or Cadogan and Callender. The paper by Burkard is quite interesting. In it, many iterated sequences are discussed.

REFERENCES

[1] J. B. Roberts, *Amer. Math. Monthly* 79 (1972) Problem E 2356.

[2] C. C. Cadogan and B. A. Callender , A problem on positive integers, *N. Z. Math. Mag.* 11 (1974) 87–91, 94.

[3] R. E. Burkard, Itierte zahlentheoretische Funktionen, *Math.-Phys. Semesterberichte* 19 (1971) 235–253.

5 Influence of Computing

Every set of 7 consecutive integers greater than 36 contains a multiple of a prime greater than 41.

Lehmer has the following to say.

> We should regard the digital computer system as an instrument to assist the exploratory mind of the number theorist in investigating the global and local properties of his universe, the natural numbers and their finite extensions.

REFERENCES

D. H. Lehmer, "The influence of computing on research in number theory" in *The Influence of Computing on Mathematical Research and Education,* Proc. Symp. Appl. Math. vol. 20, Amer. Math. Soc., Providence, 1974, pp. 3–12.

6 Fibonacci Numbers of the Form $k^2 + 1$

The nth Fibonacci number u_n ($u_1 = u_2 = 1$) is 1 more than a square for exactly 7 values of n and they are $\pm 1, 2, \pm 3, \pm 5$.

REFERENCES

H. C. Williams, On Fibonacci numbers of the form $k^2 + 1$, *Fib. Quarterly* 13 (1975) 213–14.

7 A Conjecture of Ramanujan

Problem 465 in the *J. Ind. Math. Soc.* [5 (1913) 120] was posed by Ramanujan. He observed that the equation $2^{n+2} - 7 = x^2$ has integer solutions for n, x when $n = 1, 2, 3, 5, 13$ and conjectured there were no others.

The problem seems not to have been settled until 1958 when Skolem proved the conjecture at the University of Notre Dame Number Theory Seminar in the spring of that year.

In 1959 Skolem, Chowla, and Lewis published that solution. They proved the theorem: *if* $a_0 = a_1 = 1, a_n = a_{n-1} - 2a_{n-2}$, *for* $n > 1$ *then*

i. $a_{n-1}^2 = 1$ *exactly when* $2^{n+2} - 7 = x^2$ *has a solution;*

ii. *any given integer appears in the sequence* $\{a_n\}$ *at most three times.*

Showing that 1 occurs exactly twice and -1 exactly three times proves the conjecture.

They also proved the following generalization. If A is an integer and $A \not\equiv 1 \pmod 8$, then $2^n + A = x^2$ has at most one integer solution (n, x). If there is a solution, then $0 \leq n \leq 2$.

In 1960 Nagell also gave a proof of Ramanujan's conjecture. The equation $2^{n+2} - 7 = x^2$ is sometimes referred to as the *Ramanujan–Nagell equation.*

In a paper on differential algebra, Mead said,

> However, it may be of some interest to note that the same problem (the equation of Ramanujan–Nagell) which has attracted a fair amount of theoretical attention over the years, also arose in the study of error-correcting codes, and has now reappeared in a problem in differential algebra. One wonders whether there is perhaps something fundamental about Ramanujan's problem, as well as when and where it may arise again.

As always, generalizations of the Ramanujan-Nagell equation have been studied. One such equation is $y^2 + 7^m = 2^n$. Independently, Toyoizumi and Tanahashi have shown that the only solutions of this equation are

$$(y, m, n) = (1, 1, 3), (3, 1, 4), (5, 1, 5), (11, 1, 7), (13, 3, 9), (181, 1, 15).$$

Brown has shown the equation $x^2 + 3 = 7^n$ only has $(n, x) = (1, \pm 2)$ as solutions. This is a special case of a theorem of Nagell who proved

[*Norsk. Mat. For. Skr.* (1) (1923) No.13 (1924)] that the equation $x^2 + 3 = y^n$ has no integral solutions with $n > 1$.

Several authors, see Cohen, have shown that the only solution to the equation $x^2 + 11 = 3^n$ is $(n, x) = (3, 4)$.

REFERENCES

[1] T. Skolem, S. Chowla, and D. J. Lewis, The Diophantine equation $2^{n+2} - 7 = x^2$ and related problems, *Proc. Amer. Math. Soc.* 10 (1959) 663–69.

[2] D. G. Mead, The equation of Ramanujan–Nagell and $[y^2]$, *Proc. Amer. Math. Soc.* 41 (1973) 333–341.

[3] M. Toyoizumi, On the Diophantine equation $y^2 + D^m = 2^n$, *Comment. Math. U. St. Paul* 27 (1978/79) 105–111.

[4] E. L. Cohen, Sur une équation diophantiennede Ljunggren, *Am. Sci. Math. Québec* 2 (1978) 109-112

Integer 8

1 Square Identities

A very old identity is

$$(x_0^2 + x_1^2)(y_0^2 + y_1^2) = (x_0 y_0 - x_1 y_1)^2 + (x_0 y_1 + x_1 y_0)^2.$$

This identity shows that the product of two sums of two squares is itself a sum of two squares. This identity is one formulation of the fact that the modulus of the product of two complex numbers is equal to the product of their moduli:

$$|(x_0 + ix_1)(y_0 + iy_1)| = |x_0 + ix_1||y_0 + iy_1|.$$

On May 4, 1748 Euler, then in Berlin, wrote a letter to Goldbach in St. Petersburgh in which he wrote the identity

$$(x_0^2 + x_1^2 + x_2^2 + x_3^2)(y_0^2 + y_1^2 + y_2^2 + y_3^2) = z_0^2 + z_1^2 + z_2^2 + z_3^2,$$

where

$$z_0 = x_0 y_0 - x_1 y_1 - x_2 y_2 - x_3 y_3$$

$$z_1 = x_0 y_1 + x_1 y_0 + x_2 y_3 - x_3 y_2$$

$$z_2 = x_0y_2 + x_2y_0 + x_3y_1 - x_1y_3$$

$$z_3 = x_0y_3 + x_3y_0 + x_1y_2 - x_2y_1.$$

This identity was used by Euler to prove that every integer is the sum of four squares.

In 1822 an eight square identity was published by Degen.

Cayley gave a modified form of such an identity in 1845. Following Dickson and writing $jk = x_jy_k - x_ky_j$ for $jk \neq 0$ and $0k = x_0y_k + x_ky_0$, Cayley's identity reads

$$(x_0^2 + \cdots + x_7^2)(y_0^2 + \cdots + y_7^2) = z_0^2 + \cdots + z_7^2,$$

where

$$z_0 = x_0y_0 - x_1y_1 - x_2y_2 - \cdots - x_7y_7$$

$$z_1 = 01 + 23 + 45 + 76$$

$$z_2 = 01 + 31 + 46 + 57$$

$$z_3 = 03 + 12 + 47 + 65$$

$$z_4 = 04 + 51 + 62 + 73$$

$$z_5 = 05 + 14 + 36 + 72$$

$$z_6 = 06 + 17 + 54 + 53$$

$$z_7 = 07 + 25 + 34 + 61.$$

The 4 and 8 square identities arise from considerations of norms in the quaternions and in the Cayley numbers.

In 1898 Hurwitz showed that the only division algebras over the reals are of dimensions 1, 2, 4, 8. In 1923 Hurwitz proved that the only square identities of the above type are for 1, 2, 4, 8. Thus, there do not exist further square identities of the above type.

Nathanson considered the following. He let $x_1, \ldots, x_m, y_1, \ldots, y_m$ be indeterminates and asked if there are polynomials $z_1, \ldots, z_m$ in these indeterminates such that

$$(x_1^n + \cdots + x_m^n)(y_1^n + \cdots + y_m^n) = z_1^n + \cdots + z_m^n.$$

If so one says $H(m,n)$ *holds.* Then, clearly, for all m, n, $H(1,n)$, $H(m,1)$ hold and the theorem of Hurwitz above tells us that $H(m,2)$ holds if and only if $m = 1,2,4,8$. He observed that $2 \cdot 14 = 28$ is a proof that $H(3,2)$ does not hold. Similarly, he proved $H(3,3)$ and $H(2,n)$, $n > 2$, do not hold. Finally, he showed that for $m > 2$, $H(m,2n)$ holds for only finitely many n.

Adem [6] called a triple of positive integers (p,q,n) *admissible* over a field F if there is an identity of the form

$$(x_1^2 + \cdots + x_p^2)(y_1^2 + \cdots + y_q^2) = z_1^2 + \cdots + z_n^2,$$

where each z_j is a homogeneous bilinear form in the x_i and the y_j with coefficients in F.

Hurwitz had asked for all admissible triples and he and Radon, in the 1920s, settled the case where $p = n$ and F is either the real or complex field. They proved that for given n, (n,q,n) is admissible if and only if $q \le \rho(n)$, where $\rho(n) = 8a + 2^b$, when $n = 2^{4a+b} \cdot n_0$ with n_0 odd, $0 \le b \le 3$.

This has since been extended to all fields F of characteristic not zero.

Adem [7] settled the question for (p,q,n), $n \le 8$, over any field F of characteristic not zero. According to the reviewer (in *Math. Rev.*), D. B. Shapiro, "The non-trivial part is to prove that the triples $(3,5,6)$, $(4,5,7)$ and $(3,6,7)$ are not admissible over F." Later he gives further information concerning the problem. Adem has also shown the admissibility of $(16,16,32)$. The nonexistence of $(16,16,16)$ was shown by Cayley and Dickson.

Yiu [9] denoted the least such n for which (r,s,n) is admissible by $r * s$ (assuming there is such an n). By Adem's result above $r * s \le 32$. In 1986 Yiu [8] showed $r * s \ge 25$ and improved this to $r * s \ge 29$ the next year. He has also shown the nonexistence of $(12,12,20)$. The work is very difficult.

Shapiro's expository paper on this subject contains more than 90 references to the literature.

For a report on a much more general question having to do with the composition of forms see Schafer.

REFERENCES

[1] C. W. Curtis, The four and eight square problem and division algebras in *MAA Studies in Modern Algebra* Vol.2, (ed. A. A. Albert), pp. 100–125.

[2] H. S. M. Coxeter, Integral Cayley numbers, *Duke J.* 13 (1946) 561–578.

[3] O. Taussky, A determinantal identity for quaternions and a new eight square identity, *J. Math. Anal. and Appl.* 15 (1966) 162–64.

[4] H. Zassenhaus and W. Eichhorn, Herleitung von Acht-und Sechzehn-Quadrate identitäten mit Hilfe von Eigenschaften der verallgemeinerten Quaternionen und der Cayley-Dicksonschen Zahlen, *Arch. Math.* 17 (1966) 492–96.

[5] M. B. Nathanson, Products of sums of powers, *Math. Mag.* 48 (1975) 112–113.

[6] J. Adem, On the Hurwitz problem over an arbitrary field I (II), *Bol. Soc. Mat. Mexicana* 25 (1980) 29–51 and 26 (1981) 29–41.

[7] ——, On admissible triples over an arbitrary field, *Bull. Soc. Math. Belg.* 38 (1986) 33–35.

[8] P. Y. H. Yiu, Quadratic forms between spheres and the non-existence of sums of squares formulae, *Math. Proc. Camb. Phil. Soc.* 100 (1986) 493–504.

[9] ——, Sums of squares formulae with integral coefficients, *Can. Math. Bull.* 30 (1987) 318–324.

[10] D. B. Shapiro, Products of sums of two squares, *Exp. Math.* 2 (1984) 235–261.

[11] R. D. Schafer, Forms permitting composition, *Adv. in Math.* 4 (1970) 127–148.

2 Parallelizability of the Spheres

There exists a GL_m bundle over S^n, with Stiefel–Whitney class $w_n \neq 0$, only if $n = 1, 2, 4, 8$. Two consequences of this are:

1. R^n has a bilinear product operation with no zero divisors only for $n = 1,\ 2,\ 4,\ 8$;
2. The sphere S^n is parallelizable only for $n = 1, 3, 7$.

REFERENCES

[1] J. Milnor, R. Bott, On the parallelizability of the spheres, *Bull. Amer. Math. Soc.* 64 (1958) 87–89.

[2] ——, Some consequences of a theorem of Bott, *Annals of Math.* 68 (1958) 444–49.

3 Regular Lattice n-Simplices

Macdonald showed that a regular n-simplex in R^n with vertices in Z^n exists if and only if $n + 1$ is the sum of 1, 2, 4, or 8 odd squares.

See also Integer 4, Latticed Cornered Polygons.

REFERENCES

I. G. Macdonald, Regular simplices with integer vertices, *C. R. Math. Rep. Acad. Sci. Canada* 9 (1987) 189–193.

4 A Multiplicative Function

Let n and s be positive integers and put $r_s(n)$ for the number of sets of integers $x_1, x_2, \ldots, x_s$ satisfying the equation

$$x_1^2 + \cdots + x_s^2 = n.$$

The following are proved by Bateman [1].

- Putting $f_s(n) = \frac{r_s(n)}{2s}$ we find this function f_s is multiplicative if and only if $s = 1, 2, 4, 8$.
- Putting $g_s(n) = f_s(n^2)$ we find this function g_s is multiplicative precisely for $1 \leq s \leq 8$.

At a later time, Bateman [3] generalized these properties as follows.

- Put $\Theta(z) = \sum_{n=-\infty}^{\infty} z^{n^2}$ and define $r_s(n)$, for s a fixed complex number, by

$$\Theta(z)^s = 1 + \sum_{n=1}^{\infty} r_s(n) z^n,$$

  where $|z| < 1$. Further put $f_s(n) = \frac{r_s(n)}{2s}$ and $\phi_s(n) = \frac{r_s(n^2)}{2s}$.
- When s is a positive integer this $r_s(n)$ is the same as the earlier one.

The following are true.

1. $r_s(n)$ is a polynomial in s of degree n whose real roots are $\leq \frac{31}{6}$.
2. The only complex s for which $f_s(n)$ is multiplicative are $s = 0, 1, 2, 4, 8$.
3. The only real s for which $\phi_s(n)$ is multiplicative are $s = 0, 1, 2, 3, 4, 5, 6, 7, 8$.

REFERENCES

[1] P. T. Bateman, Elementary Problem E 2051, *Amer. Math. Monthly* 76 (1969) 190–191.

[2] ——, Multiplicative arithmetic functions and the representation of integers as sums of squares, *Proc. Number Theory Conf.*, Boulder, 1972, pp. 9–13.

[3] Jean Lagrange, Décomposition d'un entieren somme carres et fonction multiplicative, *Sem. Délange-Pisot-Poitou Thé. nombres Univ. Paris* (1972–73) 14, No. 1, 1/1–1/5.

5 Fibonacci Cubes

The only cubes in the Fibonacci sequence are the following:

$$u_{-6} = -8, \quad u_{-2} = -1, \quad u_0 = 0, \quad u_{-1} = u_1 = u_2 = 1, \quad u_6 = 8.$$

The only cube in the Lucas sequence is L_1 (= 1).

REFERENCES

H. London and R. Finkelstein, On Fibonacci and Lucas numbers which are perfect powers, *Fib. Quarterly* 7 (1969) 476–481, 487 (Errata *Fib. Quarterly* 8 (1970) 3, 248).

Integer 9

1 Waring for Cubes

Every integer may be written as a sum of 9 nonnegative cubes. The 9 may not, however, be changed to 8 since each of the integers 23 and 239 requires 9 cubes. These are the only integers requiring 9 cubes.

See also Integer 16, Waring for Fourth Powers, Integer 19, Waring's Problem, and Integer 239, Sums of Cubes.

REFERENCES

G. H. Hardy and E. M. Wright, *An Introduction to the Theory of Numbers,* (5th ed.), Oxford Univ. Press, Oxford, 1988, pp. 335–38.

2 Sums of Distinct Odd Primes

Every positive integer except for 1, 2, 4, 6, 9 may be written as the sum of distinct odd primes.

Some related results due to Mąkowski are the following.

1. Every integer greater than 55 is a sum of distinct $4k - 1$ primes.

2. Every integer greater than 121 is a sum of distinct $4k + 1$ primes.

3. Every integer greater than 161 is a sum of distinct $6k - 1$ primes.

4. Every integer greater than 205 is a sum of distinct $6k + 1$ primes.

Similar results due to Dressler, Mąkowski, and Parker are as follows.

5. Every integer greater than 1969 is a sum of distinct $12k + 1$ primes.

6. Every integer greater than 1349 is a sum of distinct $12k + 5$ primes.

7. Every integer greater than 1387 is a sum of distinct $12k + 7$ primes.

8. Every integer greater than 1475 is a sum of distinct $12k + 11$ primes.

All eight of these results are best possible.

The last four of these statements use the following strengthening of Bertrand's postulate, due to Molsen, in their proofs.

If $n \geq 118$, then between n and $\frac{4}{3}n$ there are primes of all of the forms $12k \pm 1, 12k \pm 5$.

See also Integer 6, Sums of Distinct Primes and Integer 33, Complete Sequences.

REFERENCES

[1] R. E. Dressler, A stronger Bertrand's postulate with an application to partitions, *Proc. Amer. Math. Soc.* 33 (1972) 226–28.

[2] A. Mąkowski, Partitions into unequal primes, *Bull. Acad. Pol. Des. Sci.* VIII (1960) 125–26.

[3] R. E. Dressler, A. Mąkowski and T. Parker, Sums of distinct primes from congruence classes modulo 12, *Math. Comp.* 28 (1974) 651–52.

[4] T. Kløve, Sums of distinct primes, *Nord. Mat. Tidskr.* 21 (1973) 138–40.

[5] K. Molsen, Zur Verallgemeinerung des Bertrandschen Postulates, *Deutsches Math.* 6 (1949) 120–22.

3 An Identity

If $a + b + c = 0$, then

$$\left(\frac{a}{b-c} + \frac{b}{c-a} + \frac{c}{a-b})(\frac{b-c}{a} + \frac{c-a}{b} + \frac{a-b}{c}\right) = 9,$$

when all quotients are defined.

This is one of those examples where a special case of a general proposition will catch the eye whereas the general proposition itself might well not be noticed at all. The result is due to C. H. Prior and was first published by J. W. L. Glaisher in 1880. MacMahon generalized this to the following.

If $\sum_{i=1}^{n} a_i^j = 0$, for all j, $1 \leq j \leq n-2$, then, if we put

$$A_k = \prod_{1 \leq i < j \leq n, i \neq k, j \neq k} (a_i - a_j),$$

$$\left(\frac{a_1}{A_1} - \frac{a_2}{A_2} + \cdots\right)\left(\frac{A_1}{a_1} - \frac{A_2}{a_2} + \cdots\right) = n^2.$$

The prior identity is obtained by putting $n = 3$. Also, if $a + b = 0$, then $(a - b)(\frac{1}{a} - \frac{1}{b}) = 2^2$.

REFERENCES

P. A. MacMahon *Collected Works,*, MIT Press, Cambridge, 1978, vol. 1, pp. 13, 20–22.

4 Public Key Cryptosystems

Every RSA public key system (based on the product of two primes) must have at least 9 messages encrypted by themselves. This is based on the following result of Blakely and Borosh.

> To every pair p, q of distinct odd primes there correspond 9 positive integers x no larger than pq such that $x^c \equiv x$

(mod pq) for every odd positive c. Therefore these 9 messages x are unconcealable in any RSA public key cryptosystem based on p, q.

REFERENCES

[1] *Applied Cryptology, Cryptographic Protocols, and Computer Security Models, Proc. of Symp. in Appl. Math.,* Amer. Math. Soc., Providence, 1983, Vol. 29, p. 47.

[2] R. Blakely and I. Borosh, Rivest–Shamir–Adelman public key cryptosystems do not always conceal messages, *Comp. Math. Appl.* 5 (1979) 169–178.

5 Partitions of a Set

We quote from an article by R. L. Graham in the *Journal of Combinatorial Theory.*

> Suppose we had a cable consisting of n indistinguishable wires with terminals at two points A and B and suppose for each wire at A it is desired to identify its mate at B. We shall assume that the only operations available for making such an identification are interconnecting sets of terminals at one end and testing for current flow in the terminals at the other end. ... we restrict ourselves further to procedures of the following type: certain connections are made at A. We then go to B and make tests and, using the test results, certain connections. We finally come back to A, disconnect the connections originally made, and perform further tests. The information now in hand should be enough to determine for each j the terminal pairs A_j and B_j of wire j.

With a cable of 6 wires at the A end join 3 of the wires and label them each with (3,); then join 2 of the remaining wires and label them with (2,); finally label the unlabelled wire with (1,). Proceed to the B end and by checking the interconnections induced by the connections at the A end label the wires

$$(3,\ ), (3,\ ), (3,\ ), (2,\ ), (2,\ ), (1,\ )$$

initially. Complete the labelling of these wires by filling in the second components as follows.

$$(3,1), (3,2), (3,3), (2,2), (2,3), (1,3).$$

Now, still at the B end, join those wires with the same second component. Return to A, disconnect the connections there and fill in the second components of the labels in accordance with the number of wires each wire is connected with via the connections at B. (Note that this can be determined without going back to B.) We now have the wires labelled at each end and furthermore the labels on the two ends of each wire agree in all cases.

This labelling depends upon our ability to choose two partitions of the set $\{1,2,3,4,5,6\}$, namely, in this case,

$$\{\{1,2,3\},\{4,5\},\{6\}\} \qquad \text{and} \qquad \{\{1,4,6\},\{2,5\},\{3\}\},$$

such that for each of the numbers from 1 to 6 the cardinalities of the parts of these two partitions which contain the numbers completely determine the numbers. Thus we have the correspondence:

$$1 \longleftrightarrow (3,3)$$

$$2 \longleftrightarrow (3,2)$$

$$3 \longleftrightarrow (3,1)$$

$$4 \longleftrightarrow (2,3)$$

$$5 \longleftrightarrow (2,2)$$

$$6 \longleftrightarrow (1,3)$$

In Graham's paper, cited below, it is shown that except for $n = 2, 5, 9$ such a pair of partitions exists whereas for these three integers no such partitions exist.

The algorithm described above is due to K. C. Knowlton.

REFERENCES

R. L. Graham, On partitions of a finite set, *J. Comb. Thy.* 1 (1966) 215–223.

Integer 10

1 1, 10, and Primes

The integer 10 is the only composite integer such that all of its positive integer divisors other than 1 are of the form $a^r + 1$, $r > 1$. Only the integers 1, 10, and primes of the form $n^2 + 1$ have the property. Since it is unknown if there are infinitely many primes of the form $n^2 + 1$ it is not known if there exist infinitely many integers with the stated property.

REFERENCES

C. D. H. Cooper, Elementary Problem E2415 *Amer. Math. Monthly* 81 (1974) 520–21.

2 Number of Relations

At most 10 distinct relations can be derived from a given relation by the formation of complements and transitive closures in succession. The bound is sharp only for sets having at least 5 elements.

If reflexive closures are allowed, the number becomes 42.

REFERENCES

R. L. Graham, D. E. Knuth and T. S. Motzkin, Complements and transitive closures, *Discrete Math.* 2 (1972) 17–29.

3 A Parlor Problem

If the first 10 positive integers are placed in a linear arrangement, show that some four of them, from left to right, form a monotone increasing or decreasing sequence. For example, if one has the arrangement

$$3, 2, 1, 6, 5, 4, 9, 8, 7, 10,$$

then $2, 6, 7, 10$ would be a monotone increasing sequence as would $1, 4, 9, 10$. In this arrangement, there is no monotone decreasing sequence of length 4.

Considering the first 9 of the integers in the above arrangement we see that if the 10 in the statement of the problem is changed to 9, the conclusion is false.

For each of the first 10 integers let j^+ be the length of the longest increasing sequence starting with j and let j^- be the length of the longest decreasing sequence starting with j. We claim that when $1 \leq i < j \leq 10$, $(i^+, i^-) \neq (j^+, j^-)$. To show this, consider the two cases $i < j$ and $j < i$ separately. If $i < j$, then $i^+ > j^+$ when i is to the left of j and $j^- > i^-$ when i is to the right of j. Similar arguments apply in the case where $i > j$. Therefore, there are exactly 10 pairs (i^+, i^-). If all of the values of the i^+ and the i^- were less than or equal to 3, then there could only be 9 such distinct pairs, contrary to the fact that there are 10 of them. Consequently, there exists a monotonic sequence of length at least 4.

The same argument may be used to show that in every linear arrangement of $mn + 1$ distinct real numbers some m of them form a monotone, left to right, increasing sequence or some n of them form a monotone, left to right, decreasing sequence.

This result seems to have appeared initially in 1935 in a paper by Erdős and Szekeres. It arose in connection with a problem in geometry. The problem had

> been suggested by Miss Esther Klein in connection with the following proposition. From 5 points of a plane of which no three lie on the same straight line it is always possible to select 4 points determining a convex quadrilateral. ... Miss Klein suggested the following more general problem. Can we find for a given n a number $N(n)$ such that from any set containing at least $N(n)$ points it is possible to select n points forming a convex polygon?

This problem was solved by Erdős and Szekeres in two ways, the second made use of the above assertion concerning the arrangements of distinct real numbers on a line.

If one puts $m(k)$ for the least number of points, no three on a line, for which some k of them form a convex polygon, then Szekeres conjectured that $m(k) = 2^{k-2}+1$. This is known to be true only for $2 \le n \le 5$. In these cases, one has

$$m(2) = 2, \quad m(3) = 3, \quad m(4) = 5, \quad m(5) = 9.$$

The best known estimate of $m(n)$ seems to be

$$2^{n-2} + 1 \le m(n) \le \binom{2n-4}{n-2}.$$

Harborth added the condition that the convex polygon contain no other point of the set in its interior and denoted the number of points needed in this case by $g(n)$. He observed that $g(2) = m(2) = 2$, $g(3) = m(3) = 3$, $g(4) = m(4) = 5$. However $g(5) = 10$, whereas $m(5) = 9$. It is not even known if $g(6)$ exists.

Harzheim called f a Z-function if it is a mapping from the set of the first n positive integers into itself such that, for all n, $f(n) \le n$. As an example, consider the Euler ϕ-function. He then asserted that every Z-function from the set of the first 2^m integers into itself is monotone increasing on some subset of those integers of cardinality $m+1$ on which the function g, defined by $g(n) = n - f(n)$, is also increasing. The 2^m may not be replaced by 2^{m-1}.

Kalmanson has given an n-dimensional version of the theorem of Erdős and Szekeres.

REFERENCES

[1] P. Erdős and G. Szekeres, A combinatorial problem in geometry, *Comp. Math.* 2 (1935) 463–470.

[2] ——, Combinatorial problems in geometry and number theory, *Relations between Combinatorics and Other Parts of Mathematics,* Proc. Symp. Pure Math. xxxiv, Amer. Math. Soc., Providence, 1979, pp. 149–162.

[3] H. Harborth, Konvexe Fünfecke in ebenen Punktmengen, *Elem. Math.* 33 (1978) 116–18.

[4] E. Harzheim, Eine kombinatorische Fragezahlentheoretischer Art, *Publ. Math. Debrecen* 14 (1967) 45–51.

[5] K. Kalmanson, On a theorem of Erdős and Szekeres, *J. Comb. Thy.* A15 (1973) 343–46.

4 Quasi-normal Sets of Integers

Let S be an infinite set of positive integers and let N_k be the number of distinct strings of k base 10 digits that occur as the k rightmost base 10 digits of numbers in S. For the sequence of primes, one finds $N_1 = 5$, $N_2 = 40$, $N_3 = 400, \ldots$.

One calls S *quasi-normal* if $N_{k+1} = 10N_k$ for all sufficiently large k. The sequence of primes is quasi-normal as is readily seen through the use of Dirichlet's theorem on primes in arithmetic progressions.

No sequence of fixed integer powers, such as squares or cubes, etc., is quasi-normal.

Examining the numerators of the convergents for the continued fraction expansion for the square root of 2, one finds $N_1 = 4$, $N_2 = 22$, $N_3 = 109$. It is not known if this sequence is quasi-normal.

REFERENCES

E. Borel, Sur une propriété arithmétique des suites illimitées d'entiers, *C. R. Acad. Sci. Paris* 233 (1951) 769–770.

5 An Exponential Diophantine Equation

Mąkowski showed that the equation $13^x - 3^y = 10$ has only the solutions $x = y = 1$ and $x = 3, y = 7$.

Chidambaraswamy, three years later, showed that only for $z = 1$ does the equation $13^x - 3^y = 10^z$ have a solution.

REFERENCES

[1] A. Mąkowski, On the equation $13^x - 3^y = 10$, *Math. Stu.* 28 (1960) and 87 (1962).

[2] J. Chidambaraswamy, On a conjecture of Mąkowski, *Math. Stu.* 31 (1963) 5–6 and (1964).

Integer 11

Sum of Distinct Primes

The largest integer not the sum of two or more distinct primes is 11.

See also Integer 6, Sum of Distinct Primes and Integer 33, Complete Sequences.

REFERENCES

H. E. Richert, Über zerfällungenin ungleiche Primzahlen, *Math. Zeit.* 52 (1941) 342–43.

Integer 12

1 Prime Factors of Fibonacci Numbers

Each term of the Fibonacci sequence $1\,(= u_1), 1, 2, 3, 5, 8, 13, 21, 34, \ldots$, with the exception of the terms numbered 1, 2, 6, 12, has a prime factor different from all prime factors appearing in earlier terms. Call such a prime factor of a Fibonacci number a *primitive* prime factor and all other prime factors *elementary* prime factors.

Yorinaga has shown that if one divides out all the elementary prime factors of the Fibonacci number u_n, $n > 5$, then any divisor N of the remaining number satisfies $u_N \equiv (N/5) \pmod{N}$, where $(N/5)$ is the Legendre symbol.

Integer 4181, Converse Numbers and Integer 2935363331541925531, Two Sequences of Composite Integers have more information on Fibonacci numbers.

REFERENCES

[1] R. D. Carmichael, On the numerical factors of the arithmetical forms $\alpha^n \pm \beta^n$, *Ann. Math.* 15 (1913) 30–70.

[2] M. Yorinaga, On a congruential property of Fibonacci numbers—Considerations and remarks, *Math. J. Okayama Univ.* 19 (1976/77) 11–17.

2 Sums of Squares

There exist exactly 12 positive integers not the sum of s squares for each of the numbers $s = 5, 6$.

For 5 they are 1, 2, 3, 4, 6, 7, 9, 10, 12, 15, 18, 33 and for 6 they are 1, 2, 3, 4, 5, 7, 8, 10, 11, 13, 16, 19.

In general for $s \geq 6$ the only positive integers not equal to a sum of s squares are the numbers $1, 2, 3, \ldots, s-1, s+1, s+2, s+4, s+5, s+7, s+10, s+13$.

Zenkin introduced the notation $N(r, s)$ for the set of integers greater than s which are not sums of precisely s positive rth powers. Then by the above

$$N(2, 6) = \{7, 8, 10, 11, 13, 16, 19\}$$

If we subtract 6 ($= s$) from each element, we get the set $Z(2) = \{1, 2, 4, 5, 7, 10, 13\}$. Hence for $s \geq 6$, $N(2, s) = \{s + z | z \in Z(2)\}$. There are similar sets definable for cubes, fourth powers, etc.

As given above, $Z(2)$ has 7 elements ≤ 13 and is "valid" for $s \geq 6$. Zenkin gives the information:

$Z(3)$ has 75 elements ≤ 149 and is valid for $s \geq 14$.

$Z(4)$ has 1321 elements ≤ 2641 and is valid for $s \geq 21$.

$Z(5)$ has 3175 elements ≤ 6261 and is valid for $s \geq 57$.

Hooley [3] reported that Littlewood once asked if for given unequal positive integers h, k there are infinitely many n for which each of n, $n + h$, $n + k$ are sums of two squares. In 1973 Hooley [4] proved that this is indeed the case. A year later he showed that for given positive integer k the number of integers not exceeding x for which n and $n + k$ are sums of two squares is, for x large enough, greater than a constant times $\frac{x}{\log x}$.

Shanks reported that Landau let $B(x)$ be the number of integers not exceeding x which are sums of two squares and that Landau proved

$$B(x) = \frac{bx}{\log x}(1 + \frac{c}{\log x} + O(\frac{1}{(\log x)^2})).$$

Shanks wrote $NH(x)$ for the number of integers not exceeding x which are not writable as a sum of two squares, i.e., are *non-hypotenuse numbers*. He showed $NH(x)$ has the same form except that the first $\log x$ is changed to $\sqrt{\log x}$. He gives values for the constants b, c in both cases.

REFERENCES

[1] G. Pall, On sums of squares, *Amer. Math. Monthly* 40 (1933) 10–18.

[2] A. A. Zenkin, A generalization of a theorem of Pall, Computations in algebra, number theory, and combinatorics, *Akad. Nauk Ukrain. SSR Inst. Kibernet. Kiev* (1980) 40–46, 86.

[3] C. Hooley, On the intervals between numbers that are sums of two squares II., *J. Number Theory* 5 (1973) 215–217.

[4] ——, On the intervals between numbers that are sums of two squares III, *J. für Reine und Angew. Math.* 267 (1974) 207–218.

[5] D. Shanks, Non-hypotenuse numbers, *Fib. Quarterly* 13 (1975) 319–321.

3 Values of $\sigma(n)$ and $\tau(n)$

The inequality $\sigma(n) < \frac{6n^{\frac{3}{2}}}{\pi^2}$ is true for all values of n other than 2, 3, 4, 6, 8, 12.

The inequality $\tau(n) < n^{\frac{2}{3}}$ is true for all values of n other than 2, 4, 6, 12.

REFERENCES

[1] U. Annapurna, Inequalities for $\sigma(n)$ and $\tau(n)$, *Math. Mag.* 45 (1972) 187–190.

[2] M. I. Israilov and I. Allikov , Some estimates for a divisor function, *Izv. Akad. Nauk USSR Ser. Fiz.-Mat.* (1980) 82–84.

4 On the Prime Factorization of Binomial Coefficients

If n, k are positive integers with $n \geq 2k$ and if we write the binomial coefficient $\binom{n}{k}$ as the product uv, where u is the product of all prime factors of the binomial coefficient which are smaller than k and v is the product of the remaining prime factors, then $u > v$ for precisely 12 pairs n, k and they are

$$(8,3),\ (9,4),\ (10,5),\ (12,5),\ (21,7),\ (21,8),\ (30,7),$$

$$(33,13),\ (33,14),\ (36,13),\ (36,17),\ (56,13).$$

If instead we take u to be the product of those prime factors less than or *equal* to k and v to be the product of the remaining prime factors, then $u > v$ holds only finitely often and 19 such cases are displayed in the paper cited.

REFERENCES

E. F. Ecklund, R. B. Eggleton, P. Erdős, and J. L. Selfridge, On the prime factorization of binomial coefficients, *J. Aust. Math. Soc.* A26 (1978) 257–269.

Integer 13

1 A Conjecture

Each of the integers 5, 6, 7, 8 are quadratic non-residues (qnr) modulo 13. Hence it is possible for a prime number p to have more than $\sqrt{p}$ consecutive integers which are qnr.

In 1932 A. Brauer showed this could not happen if $p \equiv 3 \pmod 4$. In 1971 Hudson showed the same thing for $p \equiv 1 \pmod{24}$. In Hudson's paper [1] the cases $p \equiv 5, 17 \pmod{24}$ are also eliminated. Thus, if the maximum number of consecutive qnr modulo a prime p is greater than $\sqrt{p}$ then $p \equiv 13 \pmod{24}$.

Hudson, in 1977, conjectured that 13 is the only prime p for which there are more than $\sqrt{p}$ consecutive qnr modulo p.

Earlier Hudson [2] had shown that if $r(k, m, p)$ denotes the first occurrence of m consecutive kth power residues of a $4n + 1$ prime p, then $r(k, m, p) > c \cdot \log p$ for infinitely many primes p when k is even and $m \geq 3$.

In Chowla's book, *The Riemann Hypothesis and Hilbert's Tenth Problem,* he makes the following assertions.

- If p is a prime greater than 10, then there exists a pair $x, x + 1$ of quadratic residues mod p with $x \leq 9$. (The proof is easy: if 2 is a qr take $x = 1$. Otherwise, if 5 is a qr, take $x = 4$, and if not, take $x = 9$.)

- If p is a prime greater than 77, then there exists a pair $x, x+1$ of cubic residues mod p with $x \leq 77$.
- If p is a prime greater than 1225 then there exists a pair $x, x+1$ of biquadratic residues mod p with $x \leq 1225$.

REFERENCES

[1] R. H. Hudson, On a conjecture of Issai Schur, *J. für Reine und Angew. Math.* 289 (1977) 215–220.

[2] ——, A note on Dirichlet's characters, *Math. Comp.* 27 (1973) 973–75.

[3] S. Chowla, *The Riemann Hypothesis and Hilbert's Tenth Problem,* Gordon and Breach, New York, 1965.

2 Simultaneous Quadratic Equations

Define $f_i(x) = \sum_{j=1}^{13} a_{ij} x_j^2$, $1 \leq i \leq 3$, where all the a_{ij} are integers. Now suppose the following two conditions are met.

1. For all real λ, μ, ν (not all 0) $\lambda f_1 + \mu f_2 + \nu f_3$ contains at least 11 variables explicitly.
2. There are nonsingular solutions of $f_1(x) = f_2(x) = f_3(x) = 0$ in the real and in the 2-adic fields.

Then the equations $f_1(x) = f_2(x) = f_3(x) = 0$ have a nontrivial integral solution.

The 13 is best possible since there are sets of equations in 12 variables for which the result is false.

REFERENCES

R. J. Cook, Simultaneous quadratic equations II, *Acta Arith.* 25 (1973/74) 1–5.

3 A Peculiar Property of 13

Beiler observed that if g is a primitive root of the prime p, then $h = g^{p-2}$ is another and, furthermore,

$$ind_g h \equiv ind_h g \pmod{(p-1)}.$$

Using this, he proved that 13 is the only prime having more than 2 primitive roots for which the relation displayed above holds for every pair of its primitive roots g and h.

REFERENCES

A. H. Beiler, A peculiar property of the primitive roots of 13, *Amer. Math. Monthly* 48 (1941) 185–87.

4 Three Thirteens

Let the numbers $a_1, \ldots, a_{26}$ be the numbers in the set {1, 8, 9, 22, 23, 34, 36, 48, 50, 62, 75, 83, 87, 89, 95, 97, 109, 130, 132, 134, 136, 156, 157, 158, 171, 173} and let the numbers b_i, $1 \leq i \leq 26$, be defined by $b_i = 175 - a_i$. Then

$$\sum_{i=1}^{26} a_i^j = \sum_{i=1}^{26} b_i^j \qquad \text{for } 1 \leq j \leq 14.$$

With the same a_i, b_i, we also have

$$\sum_{i=1}^{13} (b_i - a_i)^x = \sum_{i=14}^{26} (a_i - b_i)^x \qquad \text{for } x = 1, 3, 5, 7, 9, 11, 13.$$

See also Integer 3, Repeating Blocks.

REFERENCES

L. J. Lander, Three thirteens, *Math. Comp.* 27 (1973) 397.

5 The "Thirteen Spheres Controversy"

In 1694 Newton and Gregory discussed how many identical spheres could touch a sphere of the same radius. In the cannon ball packing there are 12 spheres touching a thirteenth one and Newton thought that this was the largest number possible while Gregory thought that perhaps 13 could touch a 14th of the same size. This discussion is sometimes referred to as the "thirteen spheres controversy." The first proof that Newton was correct was apparently given in 1874, independently, by C. Bender and R. Hoppe (see Conway and Sloane).

Putting $\tau = \frac{1+\sqrt{5}}{2}$, the 12 points $(0, \pm 1, \pm\tau)$, $(\pm\tau, 0, \pm 1)$, $(\pm 1, \pm\tau, 0)$ form the vertices of a regular icosahedron with edge length 2 and with each vertex at a distance of $\sqrt{2+\tau}$ from the center of the icosahedron. Consequently, if we place a sphere of radius $\frac{\sqrt{2+\tau}}{2}$ at the center and at each vertex of the icosahedron so formed, then the spheres placed at the vertices do not touch each other, but all touch the sphere placed at the center of the icosahedron. This shows that there are infinitely many ways of placing 12 identical spheres each touching another sphere identical with them. (When one thinks of the "rigidity" of the cannon ball model this seems quite surprising at first. However, it might lead one to have a little more respect for Gregory's position in the discussion with Newton.) There is enough "play" in this arrangement of the spheres to enable any given permutation of the 12 spheres to be obtained from any other permutation by a rolling of the outer spheres around the central sphere. (See Conway and Sloane for a more complete discussion.)

In discussing such problems, Melnyk, Knop, and Smith say the following.

> The problem and its equivalents and cognates, of how to arrange m points (or equivalent particles or discrete unit charges) on the surface of a three-dimensional sphere to satisfy a prescribed extremal condition, is of importance in stereochemistry . . . as well as in botany, virology, information theory and elsewhere . . .

Conway and Sloane observed that the problem of the spheres is "sometimes called Tammes' problem, after the Dutch botanist who was led to this question by studying the distribution of pores on pollen grains."

The number 12 is also called the *kissing number* for three dimensions. The kissing number is defined in the same way for other dimensions. The kissing number is known only for dimensions 1, 2, 3, 8, and 24 where it is, respectively, 2, 6, 12, 240, and 196560. However, there are bounds known in other dimensions as well. For instance, it is known that for dimension 4 the kissing number is either 24 or 25.

There is an obvious connection between the problem of 12 spheres with the problem of determining where n points should be placed on a sphere so as to maximize the minimum distance between any two of them. In 1951 Tarski (see Robinson [3]) gave a theoretical solution that showed the desired distance was an algebraic number but the method does not seem suitable for numerical calculation.

In working on the problem Robinson [4] tried to prove this conjecture.

Conjecture. *The maximum value of the minimum distance between pairs of points from $n - 1$ points on a sphere is always greater than for n points except possibly when $n = 6, 12, 24, 48, 60,$ or 120.*

Though he was unsuccessful in his original goal he did succeed in proving the following theorem.

Theorem. *If n points are placed on a sphere in such a way that each point is as near to five others as any two points are to each other, then $n = 12, 24, 48, 60,$ or 120. For any n the configuration is unique up to rotation and reflection.*

For $n = 12$ the points form the vertices of a regular icosahedron and only in this case do they possess central symmetry. The others occur in "left- and right-hand versions."

The third part of Hilbert's 18th problem reads as follows (see Milnor below): "How can one arrange most densely in space an infinite number of equal solids of given form, e.g., spheres with given radii . . ., that is, how can one so fit them together that the ratio of the filled to the unfilled space may be as great as possible?"

Continuing to quote Milnor we have: "For 2-dimensional disks this problem has been solved by Thue and Fejes Tóth who showed that the expected hexagonal (or honeycomb) packing of circular disks in the plane is the densest possible. However, the corresponding problem in 3 dimensions remains unsolved. This is a scandalous situation since the (presumably) correct answer has been known since the time of Gauss."

C. A. Rogers has said "many mathematicians believe, and all physicists know" that the densest packing is given by the cannonball configuration mentioned earlier. The ratio in that case is $\frac{\pi}{\sqrt{18}}$, which is approximately .7405. Rogers has proved the correct ratio to be bounded above by .7796. This has been improved in recent times (see Conway and Sloane) though the true value is not yet known.

See also Integer 23, Hilbert's List.

REFERENCES

[1] J. H. Conway and N. J. A. Sloane , *Sphere Packings, Lattices and Groups*, Springer, New York, 1988.

[2] T. W. Melnyk, O. Knop, and W. R. Smith, Extremal arrangements of points and unit charges on a sphere: equilibrium configurations revisited, *Can. J. Chem.* 55 (1977) 1745–1761.

[3] R. M. Robinson, Arrangement of 24 points on a sphere, *Math. Ann.* 144 (1961) 17–48.

[4] ——, Finite sets of points on a sphere with each nearest to five others, *Math. Ann.* 179 (1969) 296–318.

[5] C. A. Rogers, The packing of equal spheres, *Proc. London Math. Soc.* 8 (1958) 609–620.

[6] J. Milnor, Hilbert's problem 18: on crystalographic [sic] groups, fundamental domains, and on sphere packing in *Proc. Symp. in Pure Math.*, vol xxvii, Amer. Math. Soc., Providence, 1976, pp. 491–506.

Integer 14

A Kuratowski Result

In a topological space the maximum number of sets obtainable from a given set by applying, iteratively, the operations of closure and complementation is 14. This was proved by Kuratowski in 1922.

An example on the real line which leads to 14 distinct sets is given by:

$$[0,1) \cup (1,2) \cup \{3\} \cup \{4\} \cup \{5\} \cup ((6,7] \cap \mathbf{Q}).$$

REFERENCES

[1] C. Kuratowski, *Topologie* I, Hafner Pub. Co., New York, (1952) 23–4.

Integers 15, 54, 88

1 Universal and Almost Universal Forms

Of the positive integers up to one hundred only $\frac{1}{10}$th of them are squares; up to ten thousand only $\frac{1}{100}$th are squares; up to one million only $\frac{1}{1000}$th are squares; ... Despite this "thinning out" of the squares a famous theorem of Lagrange states that *every* positive integer is a sum of four or fewer squares (repetitions allowed).

Thus the quadratic form $x^2 + y^2 + z^2 + w^2$ "represents" all positive integers.

Given positive integers a, b, c, d, Ramanujan showed that a necessary and sufficient condition that the quadratic form $ax^2 + by^2 + cz^2 + dw^2$ be *universal* (i.e., represent all positive integers) is that it represents the first 15 positive integers.

In 1916 Ramanujan gave 55 quadruples a, b, c, d whose form is universal. One of his 55 quadruples was $1, 2, 5, 5$ and the form with these coefficients does not represent 15 and is, therefore, not universal. The remaining 54 forms are universal and, in fact, exhaust the set of all universal forms. A quadratic form of the above type is called *almost universal* if it represents all positive integers with exactly one exception. There are exactly 88 almost universal forms. An almost universal form must omit exactly one of the first 15 positive integers. The form $x^2 + 2y^2 + 7z^2 + 13w^2$ omits only 5 while the forms $2x^2 + 2y^2 + 3z^2 + 4w^2$

and $2x^2 + 3y^2 + 4z^2 + 5w^2$ omit only 1.

REFERENCES

[1] S. Ramanujan, On the expression of a number in the form $ax^2 + by^2 + cz^2 + dw^2$, *Camb. Phil. Soc.* 19 (1917) 11–21.

[2] P. R. Halmos, Note on universal forms, *Bull. Amer. Math. Soc.* 44 (1938) 141–44.

[3] G. Pall, On almost universal forms, *Bull. Amer. Math. Soc.* 46 (1940) 291.

2 Totitives

The positive integers smaller than n and prime to n form a complete system of residues modulo $\phi(n)$ only when n is prime or when $n = 15$. Thus 15 is the only composite integer with this property. (The positive integers less than and prime to 15 are 1, 2, 4, 7, 8, 11, 13, 14 and this is a complete set of residues modulo $\phi(15)$ (= 8).)

Let T_m be the set of *totitives* of m, that is, the set of positive integers not exceeding m and prime to it, and let R_m be the set of distinct residue classes modulo $\phi(m)$ with representatives in T_m. Then the following are true.

- R_m is a group under addition modulo $\phi(m)$ if and only if m is either 15 or a prime. The group is always cyclic.
- The totitives of m form a group under addition modulo $\phi(m)$ and this group is isomorphic with the multiplicative group of totitives mod m if and only if m is prime.

REFERENCES

[1] V. O. S. Olunloyo and A. D. Weiss, On the completeness of the totitives of a natural number, *Amer. Math. Monthly* 73 (1966) 490–94 (or *J. London Math. Soc.* 38 (1965) 518–522).

[2] ——, A group theoretic characterization of the primes, *Amer. Math. Monthly* 73 (1966) 858–860.

3 Waring for Fourth Powers

Every sufficiently large integer is writable as a sum of 16 fourth powers and the 16 may not be diminished. That is, there are arbitrarily large positive integers requiring 16 such powers. In fact, any integer of the form $16^h \cdot 31$ requires 16 fourth powers. Sufficiently large integers not congruent to 15 or 16 modulo 16 may be written as a sum of 14 fourth powers while sufficiently large integers not of the form $16^h \cdot k$, where k is drawn from a suitable finite set, may be written as a sum of 15 fourth powers.

See also Integer 9, Waring for Cubes, Integer 19, Waring's Problem, and Integer 239, Sums of Cubes.

REFERENCES

[1] G. H. Hardy and E. M. Wright, *An Introduction to the Theory of Numbers*, (5th ed.), Oxford Univ. Press, Oxford, 1988, pp. 335–38.

[2] H. Davenport, On Waring's problem for fourth powers, *Annals of Math.* 40 (1939) 731–747.

Integer 16

The integer 16 is the only positive integer n for which there are integers x and y such that $x^y = y^x = n$. Goldbach and Euler found that $x^y = y^x$ has the *rational* solutions

$$x = \left(1 + \frac{1}{n}\right)^n, \quad y = \left(1 + \frac{1}{n}\right)^{n+1}, \qquad n = 1, 2, \ldots.$$

In 1990 Sved proved these are the only rational solutions.

REFERENCES

Marta Sved, On the rational solutions of $x^y = y^x$, *Math. Mag.* 63 (1990) 30–33.

Integer 17

1 Sums of Relatively Prime Integers

Every integer greater than 17 is a sum of three pairwise relatively prime integers each larger than 1, but 17 is not so representable.

REFERENCES

W. Sierpiński, *250 Problems in Elementary Number Theory*, American Elsevier, New York, 1970.

2 A Ramanujan Type Congruence

Writing $p(n)$ for the number of partitions of the integer n we have

$$p(23^2 \cdot 17n + 2623) \equiv 0 \pmod{17}.$$

F. J. Dyson said "You had to be a personal friend of 17 and 23 before you could hope to discover a congruence like (13)." The congruence written above is his (13).

REFERENCES

F. J. Dyson, A walk through Ramanujan's garden in *Ramanujan Revisited* (eds. G. E. Andrews et al.), Academic Press, Boston, 1988, pp. 7–28.

3 Consecutive Integers

Pillai and, independently, Brauer showed that in every set of fewer than 17 consecutive integers there is at least one of them prime to all other members of the set but the same may not be said for 17 or more consecutive integers.

Several other results are as follows.

1. If a set of k consecutive integers contains a prime, then it contains an integer prime to all others. (Zahlen)
2. Among six consecutive sets of 17 consecutive integers there are at least five which contain an integer prime to the others. (Zahlen)
3. In 1892 Sylvester showed that every product of k consecutive integers is divisible by a prime larger than k. Erdős stated this theorem as follows. *For positive integers n, k, with $n > 2k$, $\binom{n}{k}$ has a prime factor greater than k.*
4. An amazing result of Størmer (see Lehmer below) implies that, for a given k, the product of two consecutive integers, if they're large enough, is divisible by a prime larger than k.
5. Lehmer defined $U_h(t)$ to be the greatest integer n (if any) greater than p_t, the tth prime number, such that $n(n+1)\cdots(n+h-1)$ is divisible by no prime greater than k. Clearly, when h is too large, $U_h(t)$ does not exist. (E.g., by Sylvester's result when $n = p_t$.) The least value of h for which there is no $U_h(t)$ is denoted by $f(p_t)$. Some data concerning the values of $f(n)$ follows.

t	1	2	$3-5$	$6-12$	$13-14$
$f(p_t)$	2	3	4	6	7

6. Ecklund and Eggleton have shown that among 4 consecutive integers, all greater than 11, there is at least one divisible by a prime

greater than 11.

7. Hanson has shown that $n(n+1)\cdots(n+k-1)$, $n > k$, contains a prime divisor $> \frac{3k}{2}$ with the exceptions of the following products: $3 \cdot 4$, $8 \cdot 9$, and $6 \cdot 7 \cdot 8 \cdot 9 \cdot 10$. Also, from his work on this question he shows the existence of a prime between $3n$ and $4n$ for $n > 1$.

8. Turk showed that if F is a nonconstant integral polynomial then there is a constant c, depending only on F, such that for all n, k with $1 < k < e^{c\sqrt{\log n}}$ the number of distinct prime factors in $F(n+1)F(n+2)\cdots F(n+k)$ is $\geq k$.

9. In 1980 Pomerance and Selfridge proved the following so-called "coprime mapping theorem," first conjectured by Newman.

   Theorem (Coprime Mapping Theorem). *Given N consecutive integers there is a surjective function f from the set of the first N positive integers onto the set of the given consecutive integers such that $(n, f(n)) = 1$ for all n.*

   A special case of the theorem was proved in 1963 by Daykin and Baines.

10. In 1724 Goldbach showed that the product of three consecutive integers is never a square.

11. In 1857 Liouville proved $(n+1)(n+2)\cdots(n+k)$ is never a power if one of the k integers is a prime or if $k > n-5$.

12. In 1917 Narumi showed $(n+1)\cdots(n+k)$ is never a square when $k \leq 201$.

13. Rigge, and Erdős "a few months later," in 1939, showed the product of k consecutive integers is never a square no matter how large k may be. In 1939 Erdős also proved that if a product of consecutive integers could be an lth power for some l then this could happen only finitely often.

14. In 1940 Erdős and Siegel proved that for k sufficiently large it is not possible for the product of k consecutive integers to be a power.

15. Finally, in 1975 Erdős and Selfridge showed that the product of two or more consecutive integers is never a square or higher power.

Another conjecture of Erdős, in this case still unsettled, is that when $n > 4$ the binomial coefficient $\binom{2n}{n}$ is never square-free. Though its truth is not known Sárközy showed in 1985 that it is true for suitably large n. No bound on this is at present known. However, in 1985, Goetgheluck showed that for n not a power of 2 the conjecture is true in the range

$$4 < n \leq 2^{42205184}.$$

For an up-to-date account of the more general problem where, rather than a product of consecutive integers, one deals with a product of consecutive terms in an arithmetic progression see pp. 215–223 of Tijdeman.

REFERENCES

[1] S. S. Pillai, On m consecutive integers III, *Proc. Ind. Acad. Sci. Sect. A* 13 (1941) 530–533.

[2] A. Brauer, On a property of k consecutive integers, *Bull. Amer. Math. Soc.* 47 (1941) 328–331.

[3] J.-P. Zahlen, Sur les nombres premiers a une suite d'entiers consécutifs, *Euclides,* Madrid 8 (1948) 115–121.

[4] D. H. Lehmer, The prime factors of consecutive integers, *Amer. Math. Monthly—Slaught Mem. Papers* 10 (1965) 19–20.

[5] E. F. Ecklund and R. B. Eggleton , Prime factors of consecutive integers, *Amer. Math. Monthly* 79 (1972) 1082–89.

[6] ——, A note on consecutive composite integers, *Math. Mag.* 48 (1975) 277–281.

[7] D. Hanson, On a theorem of Sylvester and Schur, *Can. Math. Bull.* 16 (1973) 195–99.

[8] J. Turk, Prime divisors of polynomials at consecutive integers, *J. für Reine und Angew. Math.* 319 (1980) 142–152.

[9] P. Erdős and J. L. Selfridge , The product of consecutive integers is never a power, *Ill. J. Math.* 19 (1975) 292–301.

[10] C. Pomerance and J. L. Selfridge, Proof of D. J. Newman's coprime mapping conjecture, *Mathematika* 27 (1980) 69–83.

[11] D. E. Daykin and M. J. Baines , Coprime mapping between sets of consecutive integers, *Mathematika* 10 (1963) 132–36.

[12] A. Sárközy, On divisors of binomial coefficients, *J. Number Theory* 20 (1985) 70–80.

[13] P. Goetgheluck, On prime divisors of binomial coefficients, *Math. Comp.* 51 (1988) 325–29.

[14] R. Tijdeman, Diophantine equations and Diophantine approximations, in *Number Theory and Applications* (ed. R. A. Mollin), Kluwer, Dordrecht, 1988, pp. 215–243.

4 The Equation $x^3 - y^2 = -17$

$$(5234)^3 - (378661)^2 = -17.$$

This is the special case $t = 1$ of the equality $x^3 - y^2 = k$, where

$$x = 6561t^{10} - 1458t^7 + 135t^4 - 4t$$

$$y = 531441t^{15} - 177147t^{12} + 26244t^9 - 11944t^6 + \frac{135}{2}t^3 - \frac{1}{2}$$

$$k = \frac{1}{4}(-81t^6 + 14t^3 - 1).$$

See also Integer 4, A Few Diophantine Equations.

REFERENCES

M. Hall, Jr., The diophantine equation $x^3 - y^2 = k$ in *Computers in Number Theory* (Atlas 2 Oxford), Academic Press, New York, 1969, pp. 173–198.

5 A Steinhaus Problem

In the 1964 English translation of his book, *One Hundred Problems in Elementary Mathematics,* Steinhaus asked the reader to find 10 numbers greater than or equal to 0 and less than 1 such that the first two are in different halves of the interval [0, 1), the first three are in different thirds of the interval, . . ., the ten are in different tenths of the interval. He then asked if the 10 could be replaced by n for n an arbitrary positive integer.

In the solutions portion of the book, he gave the following solution to the case where there are 10 numbers:

$$0.06,\ 0.55,\ 0.77,\ 0.39,\ 0.96,\ 0.28,\ 0.64,\ 0.13,\ 0.88,\ 0.48$$

He further observed that one can append to these ten numbers the numbers 0.19, 0.71, 0.35, 0.82 to give a set of 14 numbers with the requisite property and then gave an argument by Schinzel proving that for $n = 75$ no such set of numbers exists.

Finally, in a footnote, he said, “M. Warmus has proved quite recently the number $n = 17$ to be the last one for which the problem has a solution.”

In the margin of my copy of the 1959 translation into Russian of the Polish edition of Steinhaus’ book (in which the above mentioned footnote does not appear), I have written that 17 numbers can be chosen but 18 cannot and that “Steinhaus stated this on May 30, 1963.” (I do not know the source of this comment.)

That 17 **is** possible and 18 **is not** possible was included in the proof of a generalization of the problem given by Berlekamp and Graham in 1970.

Very recently Guy included this as his Example 63 in a quite amusing paper giving a number of instances illustrating his title, “The Second Strong Law of Small Numbers.”

REFERENCES

[1] H. Steinhaus, *One Hundred Problems in Elementary Number Theory*, Basic Books, New York, 1964, pp. 6–7, 61–64.

[2] E. R. Berlekamp and R. L. Graham, Irregularities in the distribution of finite sequences, *J. Number Theory* 2 (1970) 152–161.

[3] R. K. Guy, The second strong law of small numbers, *Math. Mag.* 63 (1990) 3–20.

Integer 18

1 Meeting Circles

Improving on a result of Besicovitch, Bateman and Reifenberg have independently proved that if one has a set of circles in the plane so that none of the circles contains the center of any other of the circles, then any circle of least radius among the circles meets (this includes touching) no more than 18 of the circles. Six years earlier Besicovitch had proved this with the 18 replaced by 21.

That 18 is best possible can be seen by examining the 18 circles with radii 1 and having centers, in polar coordinates,

$$\left(1, \frac{h\pi}{3}\right), \ 0 \le h \le 5, \qquad \text{and} \qquad \left(2\cos\frac{\pi}{12}, \frac{(2k+1)\pi}{12}\right), \ 0 \le k \le 11.$$

Bateman proves the result is equivalent to the proposition:

It is impossible to have 20 points in or on a circle of radius 2 such that one of the points is at the center and all mutual distances are ≥ 1.

Bateman also proves the following theorem.

A set of 7 *points in the plane whose mutual distances are at least* 1 *has diameter at least* 2 *with this value attained only by the vertices and center of a regular hexagon with side* 1.

REFERENCES

[1] P. T. Bateman, Geometrical extrema suggested by a lemma of Besicovitch, *Amer. Math. Monthly* 58 (1951) 306–314.

[2] E. R. Reifenberg, A problem on circles, *Math. Gaz.* 32 (1948) 290–92.

2 More Totitives

Recall that for n a positive integer any positive integer less than and prime to n is a *totitive* of n. The totitives of 10 are the integers $1, 3, 7, 9$ and if we add these successively to 10 we obtain $11, 13, 17, 19$; all prime. The largest integer with this property is 12 and the total collection of such integers is $\{1, 2, 4, 6, 10, 12\}$.

If one asks for integers n for which a positive integer k exists so that kn plus each totitive is prime, we find only the above integers and 18. For 18 the least such value of k is 892.

If one subtracts each of the totitives of 30 from 30, one, except for 1, always gets a prime. Is 30 the largest number for which this is true?

REFERENCES

[1] M. Hausman, H. N. Shapiro, Adding totitives, *Math. Mag.* 51 (1978) 284–88.

[2] H. G. Kopetzky, W. Schwarz, Two conjectures of B. R. Santos concerning totitives, *Math. Comp.* 33 (1979) 841–44.

3 Sums of Mixed Powers

It is known that all sufficiently large natural numbers are writable in the form

$$r_1^2 + r_2^3 + \cdots + r_s^{s+1},$$

where the r_i are natural numbers. Roth (1951) showed s could be taken to be 50, Vaughan (1970, 1971) gave $s = 30$ then $s = 26$, Thanagasalam (1980) reduced this to 22, and Brüdern (1987) showed that 18 suffices.

Scourfield proved in 1960 that if $2 \leq n_1 \leq n_2 \leq n_3 \leq \cdots$, then a necessary and sufficient condition that an s exists such that all sufficiently large integers be writable in the form

$$\sum_{i=1}^{s} x_i^{n_i}$$

is that the series $\sum_1^\infty \frac{1}{n_i}$ diverges.

Results of this kind are outgrowths of a theorem of Roth to the effect that almost all positive integers are writable as a sum of a square, a cube, and a fourth power.

Brüdern has shown that every sufficiently large integer is writable in each of the following forms: 2 squares and 6 fifth powers, 2 squares and 7 sixth powers, 1 square and 9 fourth powers, 1 square, 3 third powers, and 8 fifth powers, 8 fourth powers and 8 fifth powers.

Vaughan has shown that for given k every sufficiently large integer is the sum of a kth power, two squares, and a cube; or, as the sum of a kth power, a square, and 5 cubes.

An isolated, but almost related, result has to do with a question raised by Erdős as to whether the equation

$$1^k + 2^k + \cdots + (x-1)^k = x^k$$

is solvable in integers.

The following is quoted from Urbanowicz. "For every t there is an explicitly given number k_0 such that the equation $1^k + 2^k + \cdots + (x-1)^k = x^k$ has no integer solutions $x \geq 2$ for all $k \geq k_0$ for which the denominator of the kth Bernoulli number B_k has at most t distinct prime factors."

Two special mixed power equations are

$$13^2 + 7^3 = 2^9 \qquad \text{and} \qquad 2^7 + 17^3 = 71^2.$$

REFERENCES

[1] K. Thanagasalam, On sums of powers and a related problem, *Acta Arith.* 36 (1980) 125–141.

[2] E. J. Scourfield, A generalization of Waring's problem, *J. London Math. Soc.* 35 (1960) 98–116.

[3] K. F. Roth, A problem in additive number theory, *Proc. London Math. Soc.* 53 (1951) 381–395.

[4] R. C. Vaughan, On sums of mixed powers, *J. London Math. Soc.* 3 (1971) 677–688.

[5] J. Brüdern, Sums of squares and higher powers I, II, *J. London Math. Soc.* 35 (1987) 233–250.

[6] J. Urbanowicz, Remarks on the equation $1^k + 2^k + \cdots + (x-1)^k = x^k$, *Ned. Akad. Wet.* 91 (1988) 343–48.

Integer 19

Waring's Problem

In 1770 Waring conjectured that every integer could be written as the sum of 4 squares, as the sum of 9 cubes, as the sum of 19 fourth powers, and so on.

The general theorem that for each fixed positive integer exponent s there is an integer k such that every integer is the sum of k nonnegative sth powers was first proved by Hilbert in 1909. A simpler proof was given by Dress in 1972.

The smallest value of k that suffices for a given s is denoted by $g(s)$.

In 1770 Lagrange showed $g(2) = 4$, in 1908 Wieferich (corrected by Kempner in 1912) showed $g(3) = 9$, in 1964 Chen showed $g(5) = 37$, and for $s \geq 6$ a number of authors, including Dickson, Pillai, Chowla, and Niven, contributed to proving the following.

Put $A = [(\frac{3}{2})^s]$, $B = [(\frac{4}{3})^s]$, and $C = AB + A + B$. Then for $k \geq 6$ we have:

1. If $3^s - 2^s + 2 < (2^s - 1)A$ then $g(s) = A + 2^s - 2$.

2. If $3^s - 2^s + 2 \geq (2^s - 1)A$ then $g(s) = A + B + 2^s - 3$ when $2^s < C$ and $g(s) = A + B + 2^s - 2$ when $2^s = C$.

(For instance, with $s = 6$, we have $A = 11$, $B = 5$, $C = 71$, and

$667 = 3^6 - 2^6 + 2 < (2^6 - 1)11 = 693$; therefore, $g(6) = 11 + 2^6 - 2 = 73$.)

Thus, up until 1986, all values of $g(s)$, with the sole exception of $g(4)$, were known. Over the years the following upper bounds for $g(4)$ were obtained:

53 (Liouville 1859),

41 (Lucas 1876),

39 (Fleck 1906),

38 (Landau 1909),

37 (Wieferich 1909),

30 (Dress 1971),

22 (Thomas 1974),

21 (Balasubramanian 1979).

Other results such as that of Aulick that all integers larger than $10^{10^{88.39}}$ are expressible as a sum of 19 fourth powers were also given.

Finally, in 1986, Balasubramanian, Dress, and Deshouillers proved that $g(4) = 19$.

For a brief discussion of the 19 biquadrate theorem see Deshouillers below.

See also Integer 239, Sums of Cubes.

REFERENCES

[1] F. Dress, Théorie additive des nombres, problème de Waring et théorème de Hilbert, *Ens. Math.* 18 (1972) 175–190.

[2] G. H. Hardy and E. M. Wright, *An Introduction to the Theory of Numbers,* (5th ed.), Oxford Univ. Press, Oxford, 1988, pp. 335–338.

[3] L. E. Dickson, *History of the Theory of Numbers,* vol. 2, Chelsea, New York, 1952, Chapter 25.

[4] R. Balasubramanian, J.-M. Deshouillers and F. Dress, Problème pour

de Waring les bicarrés, *C. R. l'Acad. Sci. Paris Sér. Math.* 303 (1986) 161–63.

[5] C. Small, Waring's problem, *Math. Mag.* 50 (1977) 12–16.

[6] W. J. Ellison, Waring's problem, *Amer. Math. Monthly* 78 (1971) 10–36.

[7] B. J. Birch and R. C. Vaughan, The papers on Waring's problem, in *The Collected Papers of H. A. Heilbronn,* pp. 576–585.

[8] K. Thanigasalam, On Waring's problem, *Acta Arith.* 38 (1980) 141–155.

[9] J.-M. Deshouillers, Waring's problem and the circle method, in *Number Theory and Applications,* (ed. R. A. Mollin), Kluwer, Dordrecht, 1988, pp. 37–44.

Integers 20, 21

A Matter of Parity

As we shall see this is really not about the integers 20 and 21 but only about the fact that they have opposite parity.

Imagine two checkrooms. At both of the checkrooms, the attendants pass back the checked items in a random fashion. The first checkroom is a small one where 20 or 21 people check objects and the second is a larger one where the number of people checking objects is of the order of a thousand. The question we wish to pose is: For which checkroom is it more likely that no person retrieves their own checked items?

If there are n people checking objects, then the probability of none of them retrieving their own objects, assuming objects are returned randomly, is given by $\frac{D_n}{n!}$, where D_n is the number of arrangements of the first n positive integers with the integer j not in the jth position (for all $j, 1 \leq j \leq n$). Call such arrangements *derangements.*

For $n \geq 3$, the derangements with 1 in the ith position are of two types:

1. There are exactly D_{n-2} derangements with i in the first position, and

2. There are exactly D_{n-1} derangements with i not in the first position.

Taking into account that there are $n-1$ possible i we see that

$$D_n = (n-1)(D_{n-1} + D_{n-2}).$$

Noting that $D_1 = 0$, $D_2 = 1$ we may use this recursion first to find $D_n - nD_{n-1} = -(D_{n-1} - (n-1)D_{n-2})$ and then $D_n - nD_{n-1} = (-1)^n$.
Dividing by $n!$ and rearranging yields

$$\frac{D_n}{n!} = \frac{D_{n-1}}{(n-1)!} + \frac{(-1)^n}{n!}.$$

From this we immediately see that

$$\frac{D_n}{n!} = \frac{1}{2!} - \frac{1}{3!} + \frac{1}{4!} - \cdots + \frac{(-1)^n}{n!}.$$

We now see that

$$\left|\frac{D_n}{n!} - \frac{1}{e}\right| < \frac{1}{(n+1)!}$$

and that the quantity inside the absolute value sign is positive for n even and negative otherwise.

We thus realize that our question cannot be answered until we know exactly how many people checked objects at the smaller of the two checkrooms. If there were 20, then at the larger checkroom it is more likely that at least one person retrieves their own objects, whereas if 21 had checked objects at the smaller checkroom, then for that checkroom it is more likely that at least one person retrieves their own objects.

The first few values of D_n, for $n = 0, 1, \ldots, 9$, are: $0, 0, 1, 2, 9, 44, 265$, 1854, 14833, 133496. This quantity is sometimes called a subfactorial and denoted by $!n$. An amusing observation (given by Madachy [2, p. 167] in the alternative notation) is:

$$148349 = D_1 + D_4 + D_8 + D_3 + D_4 + D_9.$$

Using the theorem of Schwartz, see Integer 4, Factorials, there are only finitely many integers of this type.

In the literature one sometimes finds $!n$ standing for the sum of the ordinary factorials of the integers from 0 to $n-1$ inclusive. Kurepa used this notation and, with it, formulated the following.

Kurepa Conjecture. $(!n, n!) = 2$ *for* $n > 1$.

(We use (a, b) for the greatest common divisor of a and b.) For information concerning these "left factorials" and their extension into the complex plane see Section B64 of *Reviews in Number Theory 1973–1983*.

REFERENCES

[1] C. L. Liu, *Introduction to Combinatorial Mathematics,* McGraw-Hill, New York, 1968.

[2] J. S. Madachy, *Mathematics on Vacation,* Scribners, New York, 1966.

Integer 21

Squared Squares

In 1903 Dehn showed that if a rectangle could be dissected into a finite number of pairwise distinct squares, then the sides of the rectangle must be commensurable.

In 1925 Morón showed how to cut a 32×33 rectangle into a finite number of distinct squares. The number of squares used in his dissection was 9 and it is now known that no number fewer than 9 will suffice. It is also now known that there are exactly two dissections of this rectangle into 9 squares.

In 1938 Hugo Steinhaus stated that it was not known if a square could be cut into a finite number of pairwise distinct squares.

In 1939 R. Sprague showed every square may be cut into 55 pairwise distinct squares. He used a method, first outlined by Stöhr, that made use of two distinct squarings of the same rectangle. In 1940 he also showed that if a rectangle had sides which were commensurable, then it was squareable. This, with Dehn's result mentioned above, gives the theorem below.

Theorem. *A rectangle is squareable if and only if its sides are commensurable.*

In 1948 F. G. Willcocks published in *The Fairy Chess Review* a cutting of a square into 24 pairwise distinct squares. This dissection was long used on the cover of the *Journal of Combinatorial Theory.*

In the late forties and fifties the mathematicians Bouwkamp, Brookes, Smith, Stone, and Tutte gave a number of constructions of squared squares and rectangles and gave general methods for such constructions based on the Kirchhoff laws in electrical circuit theory.

In 1962 A. J. W. Duijestijn showed, making use of the computer, that no square could be cut into fewer than 20 pairwise distinct squares. On "the night of March 22, 1978" Duijestijn showed that the least number of squares in a squaring of a square is 21 and exhibited such a covering of a square. This squaring is now used on the cover of the *Journal of Combinatorial Theory.*

In a review Bouwkamp observed:

> There are 154490 simple perfect squared rectangles of order less than 19. There is no perfect squared square of order less than 20. There will be about 80 million simple perfect squared rectangles of order less than 24.

A *simple* squared square (as used in the above quotation) is a squared square in which no subrectangle of the square is squared by a subset of the squares used in the squaring. A squared square that is not simple is called *compound.*

In 1982 Duijestijn, Federico, and Leeuw showed that there exists a unique compound perfect squared square of order 24 but none of any smaller order.

Brooks, Smith, Stone, and Tutte observed that any subdividing of an equilateral triangle into two or more, but finitely many, equilateral triangles must have at least two of the latter congruent but that the same is not true of isosceles triangles. They also mentioned other generalizations of the "squaring" problem.

For a recent discussion on tiling squares with triangles see Guy below.

See also Integers 2, 3, Cubing Cubes and Integer 5, The Banach–Tarski Paradox.

REFERENCES

[1] A. J. W. Duijestijn, Simple perfect squared square of lowest order, *J.*

Comb. Thy. B 25 (1978) 240–43.

[2] W. T. Tutte, "Squaring the square" in *The 2nd Scientific American Book of Puzzles and Diversions,* Simon and Schuster, New York, 1961, pp. 186–209.

[3] N. D. Kazarinoff and R. Weitzenkamp , Squaring rectangles and squares, *Amer. Math. Monthly* 80 (1973).

[4] C. J. Bouwkamp, *Math. Rev.* 35 (1968) #5346.

[5] A. J. W. Duijestijn, P. J. Federico, and P. Leeuw, Compound perfect squares, *Amer. Math. Monthly* 89 (1982) 15–32.

[6] R. L. Brookes, C. A. B. Smith, A. H. Stone, and W. T. Tutte, The dissection of rectangles into squares, *Duke J.* 7 (1940) 312–340.

[7] R. K. Guy, "Tiling the square with rational triangles" in *Number Theory and Applications* (ed. R. A. Mollin), Kluwer, Dordrecht, 1988, pp. 45–101.

Integer 22

A Special Congruence

The only composite integers satisfying the congruence

$$n\sigma(n) \equiv 2 \pmod{\phi(n)}$$

are 4, 6, and 22.

Any composite $n > 4$ satisfying

$$\phi(n)\tau(n) \equiv -2 \pmod{n}$$

must have at least four distinct prime factors. ($\tau(n)$ is the number of divisors of n.)

REFERENCES

M. V. Subbarao, On two congruences for primality, *Pac. J. Math.* 52 (1974) 261–68.

Integer 23

1 LCM of Binomial Coefficients

I. S. Williams, in the *Bulletin of the Australian Mathematical Society,* observed

> Recently Kurt Mahler asked: for which natural numbers N is the least common multiple of all the binomial coefficients $\binom{N}{k}$ the product of the primes less than or equal to N.

The answer is that it is true only for the integers 2, 11, 23.

In the proof it is shown that the lcm of the first $N+1$ natural numbers divided by $N+1$ is equal to the lcm in question and from this the result is derived. This preliminary result was proposed by Montgomery as a problem in the *American Mathematical Monthly*.

From the above we see that only for $n = 2, 11, 23$ is the least common multiple of $1, 2, \ldots, n+1$ equal to the product of the primes not exceeding n.

A similar question for gcd's is answered by the following.

If p is a prime and r is the highest power of p dividing N, then

$$\gcd\left\{\binom{N}{k} | 1 \leq k \leq N,\ (k,p) = 1\right\} = p^r.$$

REFERENCES

[1] I. S. Williams, On a problem of Kurt Mahler concerning binomial coefficients, *Bull. Aust. Math. Soc.* 14 (1976) 299–302.

[2] P. L. Montgomery, Elem. Prob. E 2686, *Amer. Math. Monthly* 84 (1977) 820, 86 (1979) 131.

[3] J. Albree, The gcd of certain binomial coefficients, *Math. Mag.* 45 (1972) 259–261.

2 Pseudoprimes

A composite integer n is a *pseudoprime* if $2^n \equiv 2 \pmod{n}$. The smallest $k > 1$ such that kn is a pseudoprime for some pseudoprime n is $k = 23$. (In the older literature one sometimes finds pseudoprimes called *Poulet numbers.*)

The numbers 341, 561, and 161038 are pseudoprimes as are all composite Fermat numbers $2^{2^n} + 1$. Lehmer has shown that the number 161038 $(= 2 \cdot 73 \cdot 1103)$ is the smallest even pseudoprime. In 1951 Beeger showed the existence of infinitely many pseudoprimes. In fact, if n is an odd pseudoprime, then $2^n - 1$ is a larger one. He also proved the existence of infinitely many even pseudoprimes.

McDaniel has given two other even pseudoprimes that are products of exactly three primes. They are $2 \cdot 178481 \cdot 154565233$ and $2 \cdot 1087 \cdot 164511353$. He also says "We believe, but have not shown, that $N = 465794$ is the smallest integer such that $2^N - 2$ is a pseudoprime."

Another interesting family of pseudoprimes is given in the following. If $n_1 < \cdots < n_k < 2^n$, then $F_{n_1} F_{n_2} \cdots F_{n_k}$ is a pseudoprime, where $F_m = 2^{2^m} + 1$ is the mth Fermat number.

In 1950 Erdős showed that $\frac{2^{2p}-1}{3}$ is a pseudoprime for all prime p greater than 3. In 1964 Rotkiewicz [4] showed that $\frac{2^{2p}+1}{5}$ is a pseudoprime for all prime p greater than 5.

Rotkiewicz [5] has shown that if p, q are distinct primes, then pq is a pseudoprime if and only if $(2^p - 1)(2^q - 1)$ is a pseudoprime. He also shows that for every prime p other than 2, 3, 5, 7, 13, and for none of these, there is a prime q for which pq is a pseudoprime. He has also shown that the only pseudoprime squares smaller than 10^{12} are 1093^2 and 3511^2. (See Integer 1093, Fermat's Conjecture, for another appearance of 1093 and 3511.)

Sometimes the word "pseudoprime" is given a slightly different meaning. For instance Hardy and Wright say a composite n is a **pseudoprime with respect to** a if

$$a^{n-1} \equiv 1 \pmod{n}.$$

In this sense, Cipolla showed, in 1903, that there are infinitely many pseudoprimes with respect to every $a > 1$.

Those n that are pseudoprimes with respect to all a prime to n are called *Carmichael numbers.* See Integer 561, Carmichael Numbers.

Somer calls N a *d-pseudoprime* if N is composite and there exists an a for which the exponent modulo N is $\frac{N-1}{d}$. For fixed d he shows the existence of infinitely many N such that N is a d-pseudoprime.

He gives the following table of all d-pseudoprimes for $1 \leq d \leq 8$.

d	N	Is N a Carmichael number?	N is d-pseudoprime to bases:
1–4	none		
5	6601	yes	
6	25	no	7, 18
7	15	no	4, 11
7	561	yes	
8	49	no	19, 31

In the collections *Reviews in Number Theory* by LeVeque (1940–1972) and by Guy (1973–1983) we find 55, respectively, 28 papers devoted to pseudoprimes. Rotkiewicz is an author of 30 papers in the LeVeque book and 12 in Guy's book. In 1972 he also wrote a book of 169 pages titled *Pseudoprime Numbers and their Generalizations.* In the review of Rotkiewicz's book, Lehmer said

> This unique book gives all that is known about a mysterious infinite set of positive integers called pseudoprimes and their relatives. ... The first 60 pages of this work is devoted to the history of the subject beginning around 500 B.C., when the Chinese opined that pseudoprimes do not exist, through the discovery of the first pseudoprime, 341, in 1819, to the year 1969.

REFERENCES

[1] A. Rotkiewicz, Sur les nombres naturel n et k tels que les nombres n et kn sont à la fois pseudopremiers, *Atti Accad. Naz. Lincei Rend. Cl. Sci. Fiz. Mat. Natur.* 8 (36) (1964) 816–18.

[2] N. G. W. H. Beeger, On even numbers m dividing $2^m - 2$, *Amer. Math. Monthly* 58 (1951) 553–55.

[3] W. L. McDaniel, Some pseudoprimes and related numbers having special forms, *Math. Comp.* 53 (1989) 407–409.

[4] A. Rotkiewicz, Sur les formules donnant des nombres pseudopremiers, *Colloq. Math.* 12 (1964) 69–72.

[5] ——, Sur les nombres pseudopremiers de la forme $M_p M_q$, *Elem. Math.* 20 (1965) 108–9.

[6] J. B. Roberts, *Elementary Number Theory,* MIT Press, Boston, 1977, pp. 66, 87s–88s.

[7] G. H. Hardy and E. M. Wright, *An Introduction to the Theory of Numbers* (5th ed.), Oxford Univ. Press, Oxford, 1988, pp. 72, 81.

[8] L. Somer, "On Fermat *d*-pseudoprimes" in *Number Theory* (eds. J.-M. Dekoninck, C. Levesque), de Gruyter, New York, 1989, pp. 841–860.

[9] A. Rotkiewicz, *Pseudoprime Numbers and their Generalizations,* Stu. Assoc. Fac. Sci. Univ. Novi Sad, Novi Sad, 1972.

[10] D. H. Lehmer, *Math. Rev.* 48 (1974) #8373.

3 Filling Boxes

We consider the problem of exactly filling a rectangular box with a finite number of smaller boxes. A collection of n boxes is called *completely incongruent* if all of the $3n$ dimensions are different.

A perfectly decomposed rectangular solid (PDRS) , is a subdivision of a box into n completely incongruent boxes. The number n is called the *order* of the PDRS.

There exists a PDRS of order 23 but of no smaller order.

There are 56 essentially distinct PDRS's with order 23. See also Integers 2, 3, Cubing Cubes and Integer 47, Cutting a Cube into Cubes.

REFERENCES

W. H. Cutler, Subdividing a box into completely incongruent boxes, *J. Rec. Math.* 12 (1979/80) 104–111.

4 Hilbert's List

At the second International Congress of Mathematicians, held August 6–12 in Paris in 1900, David Hilbert, in an invited address, listed a series of 23 problems whose solutions he believed would be of such a nature that the mathematics of the coming century would be significantly influenced by them. He began his talk (from the translation by Newson):

> Who of us would not be glad to lift the veil behind which the future lies hidden, to cast a glance at the next advances of our science and at the secrets of its development during future centuries.

A little later as he discussed the role of problems in the development of mathematics, he said,

> We can ask whether there are general criteria which mark a good mathematical problem. An old French mathematician said "A mathematical theory is not to be considered complete until you have made it so clear that you can explain it to the first man whom you meet in the street." This clearness and ease of comprehension, here insisted on for a mathematical theory, I should still more demand for a mathematical problem if it is to be perfect; for what is clear and easily comprehended attracts, the complicated repels us. Moreover a mathematical problem should be difficult in order to entice us, yet not completely inaccessible, lest it mock at our efforts. It should be to us a guidepost on the mazy paths to hidden truths, and ultimately a reminder of our pleasure in the successful solution.

After further discussion of a general nature, Hilbert proceeded to the outline of the 23 problems.

In connection with the above quote we also recall the following 1947 remarks of John von Neumann, made in an article titled "The Mathematician."

> One expects a mathematical theorem or a mathematical theory not only to describe and to classify in a simple and elegant way numerous and a priori disparate special cases. One also expects "elegance" in its "architectural" structural makeup. Ease in stating the problem, great difficulty in getting hold of it and in all attempts at approaching it, then again some very surprising twist by which the approach, or some part of the approach, becomes easy, etc. Also, if the deductions are lengthy or complicated, there should be some simple general principle involved, which "explains" the complications and detours, reduces the apparent arbitrariness to a few simple guiding motivations, etc. These criteria are clearly those of any creative art ...

In his excellent exposition (*Amer. Math. Monthly* 80 (1973)) of the negative solution of Hilbert's tenth problem, Davis had the following to say.

> When a long outstanding problem is finally solved, every mathematician would like to share in the pleasure of discovery by following for himself what has been done. But too often he is stymied by the abstruseness of so much of contemporary mathematics. The recent negative solution to Hilbert's tenth problem given by Matijasevič ... is a happy counterexample. In this article, a complete account of this solution is given; the only knowledge a reader needs to follow the argument is a little number theory: specifically basic information about divisibility of positive integers and linear congruences.

An interesting sidelight is the following. A mathematician named Hermann read in a paper by Davis [4] that if the Diophantine equation

$$9(x^2 + 7y^2)^2 - 7(u^2 + 7v^2)^2 = 2$$

had no nontrivial solutions (i.e., solutions different from $x = \pm 1, y = 0, u = \pm 1, v = 0$), then Hilbert's tenth problem would be unsolvable in the sense of recursive number theory. In working on the question of the solvability of this equation, Hermann showed the existence of a nontrivial solution. Thus his work had no bearing on the solution to Hilbert's tenth problem; it was solved by Matijasevič while Hermann's paper was going to press.

In 1973 a book of letters written by Hermann Minkowski to David Hilbert was published and in a letter written on January 5, 1900 Minkowski said,

> What would have the greatest impact would be an attempt to give a preview of the future, i.e., a sketch of the problems with which the future mathematicians should occupy themselves. In this way you could perhaps make sure that people would talk about your lecture for decades in the future.
>
> —Translated by H. Koch

See also Integers 2, 3, Equidecomposability; Integer 5, Primes as Range of a Polynomial and Integer 13, The "Thirteen Spheres Controversy."

REFERENCES

[1] Mathematical Problems, translated by M. W. Newson, *Bull. Amer. Math. Soc.* 8 (1902) 437–479; reprinted in *Mathematical Developments Arising from Hilbert Problems,* Proc. of the Symp. in Pure Math. xxviii, 2 vols., Amer. Math. Soc., Providence, 1976.

[2] J. Von Neumann, *The World of Mathematics,* (ed. J. R. Newman), Simon and Schuster, New York, 1956, pp. 2053–2068.

[3] M. Davis, Hilbert's tenth problem is unsolvable, *Amer. Math. Monthly* 80 (1973) 233–269.

[4] ——, One equation to rule them all, The Rand Corp. Memorandum RM-5494-PR, 1968.

[5] O. Hermann, "A non-trivial solution of the Diophantine equation $9(x^2 + 7y^2)^2 - 7(u^2 + 7v^2)^2 = 2$," in *Computers in Number Theory*

(eds. A. O. L. Atkin, B. J. Birch), Academic Press, New York, 1971, pp. 207–212.

[6] *Hermann Minkowski: Briefe an David Hilbert* (eds. L. Rüdenberg, H. Zassenhaus), Springer, New York, 1973, pp. 119–121.

[7] H. Koch, *Zbl.* 66 2 #01018, *Math. Rev.* 57 #5645.

5 Birthday Problems

Assume that all dates are equally likely for birthdays and that people are chosen randomly from the population. (We exclude February 29 from consideration.) Then we may ask the following two questions.

1. What is the minimum number of people necessary so that the probability of one of them having your birthday is $> \frac{1}{2}$?
2. What is the minimum number of people necessary so that the probability of **some** pair of them having the same birthday is $> \frac{1}{2}$?

Following Mosteller, we call the first problem the *birthmate problem* and the second the *birthday problem.*

In the **birthmate** problem, the probability that in a collection of n people some one of them, at least, has your birthday is

$$Q_n = 1 - \left(1 - \frac{1}{365}\right)^n$$

while in the **birthday** problem the probability that in a collection of r people some two of them have the same birthday is

$$P_r = 1 - \frac{365 \cdot 364 \cdots (365 - (r-1))}{365^r}.$$

Asking for the least such n, r with probability exceeding $\frac{1}{2}$ yields, in the two cases, $n = 253$ and $r = 23$. It is interesting to note the following:

the number of ways of getting a birthmate $= n$

the number of ways of getting a match $= \binom{r}{2}$.

One might think, intuitively, that $n \approx \binom{r}{2}$ for the least n, r in question.
Putting $r = 23$ we find

$$\binom{r}{2} = \binom{23}{2} = \frac{23 \cdot 22}{2} = 253.$$

More generally, setting $Q_n = P_r$ we find, after some simplifications,

$$\left(1 - \frac{1}{365}\right)^n = \left(1 - \frac{1}{365}\right)\left(1 - \frac{2}{365}\right)\cdots\left(1 - \frac{r-1}{365}\right).$$

The left side is

$$1 - \frac{n}{365} + \frac{\binom{n}{2}}{365^2} - \cdots$$

and the right side is

$$1 - \frac{1 + 2 + \cdots + (r-1)}{365} + \cdots.$$

Neglecting higher order terms we see that $n \approx \frac{r(r-1)}{2}$.
For a more careful analysis along these lines, see Mosteller below. He gave the following table.

target probability	n	Q_n	r	P_r
.01	4	.01091	4	.01636
.1	39	.10147	10	.11695
.3	131	.30190	17	.31501
.5	253	.5048	23	.50730
.7	439	.70013	30	.70632
.9	840	.90019	57	.99102
.99	1679	.99001	57	.99012

An interesting variant of these problems was given by Nymann. For n, k positive integers, $k \leq n$, let $P(n, k)$ be the probability that in a group of n ordered people one of the first k people has a birthday matching some other one of the n people.

It is not difficult to see

$$P(n,k) = 1 - (1 - P(k,k))\left(1 - \frac{k}{365}\right)^{n-k}.$$

Note that $P(r,r) = P_r$ and $P(n+1,1) = Q_n$.

In a problem in *The Mathematical Gazette,* Slomson asked how many people were necessary in order that the probability of some 3 of them having the same birthday is greater than $\frac{1}{2}$. The result is that 88 people suffice and if one changes the 3 to 4, then 187 is the appropriate number.

In the solution to the problem [3], a very nice method of carrying out the calculations was given and was attributed to K. Thomas. It goes as follows.

Let P_n be the probability that in a collection of n randomly chosen people no three of them have the same birthday. Clearly, this is zero if $n > 730$ so we assume this is not the case. By simple combinatorial reasoning, one sees that

$$P_n = \frac{n!c_n}{365^n},$$

where c_n is the coefficient of x^n in the expansion of

$$f(x) = \left(1 + x + \frac{x^2}{2}\right)^{365}.$$

Thus $c_n = f^n(0)/n!$ and, therefore,

$$P_n = \frac{f^n(0)}{365^n}.$$

By calculating the derivative of f we see

$$\left(1 + x + \frac{x^2}{2}\right) f'(x) = 365(1+x)f(x).$$

Differentiating $n+1$ times by using Leibniz' rule yields

$$\left(1 + x + \frac{x^2}{2}\right) f^{(n+2)}(x) + (1+x)(n+1)f^{(n+1)}(x) + \binom{n+1}{2} f^{(n)}(x) =$$

$$365(1+x)f^{(n+1)}(x)+365(n+1)f^{(n)}(x).$$

Putting $n = 0$, dividing by 365^{n+2}, replacing the derivatives by their equals in terms of the P_n, and rearranging terms gives

$$P_{n+2}-P_{n+1}=\frac{n+1}{365}\left\{(P_n-P_{n-1})-\frac{n}{730}P_n\right\}.$$

Putting $Q_k = P_k - P_{k+1}$ this last equation becomes

$$Q_{n+1}=\frac{n+1}{365}\left(\frac{n}{730}P_n-Q_n\right).$$

This equation along with $P_{n+1} = P_n - Q_n$, $P_0 = 1$, $Q_0 = 0$ enables one to calculate (on a computer) the greatest value of n for which P_n is less than $\frac{1}{2}$.

REFERENCES

[1] F. Mosteller, Understanding the birthday problem, *The Math. Teacher* (1962) 322–25.

[2] J. E. Nymann, Another generalization of the birthday problem, *Math. Mag.* 48 (1975) 46–7.

[3] A. Slomson, Problem 70B, *Math. Gaz.* 70 (1986) 52, 228–29.

6 The τ Function

As is discussed in some greater detail in Integer 63001, The Ramanujan τ Function, we define $\tau(n)$ by

$$x\left((1-x)(1-x^2)(1-x^3)\cdots\right)^{24}=\sum_{n=1}^{\infty}\tau(n)x^n.$$

We discuss here only one of the many interesting congruence properties of this function, that having to do with 23. (See, however, Integer 24, $\sigma(24n-1)$.) Ramanujan showed that $\tau(n) \equiv 0\ (mod\ 23)$ if $(\frac{n}{23}) = -1$. This has been extended, for primes $p \neq 23$, to

$$\tau(p) \equiv \begin{cases} 0 \pmod{23} & \text{if } (\frac{p}{23}) = -1; \\ 2 \pmod{23} & \text{if } p = u^2 + 23v^2 \text{ with } u \neq 0; \\ -1 \pmod{23} & \text{for other } p \neq 23. \end{cases}$$

Swinnerton-Dyer observed, "One reason for improving Ramanujan's congruences was the hope of proving by congruence considerations Lehmer's conjecture that $\tau(n)$ never vanishes." He also asked, in contemplation of a considerably expanded set of congruences of this type, "can these congruences be further improved, are there congruences modulo powers of other primes, and can the results be fitted into a systematic pattern?" The study is far from elementary.

It is amusing to note that later, in referring to one of the small parts of the enquiry, he observed, "it is unlikely to require fundamentally new ideas; it can be recommended only to a doctoral student desperate for a soluble but not-yet-solved problem."

Before leaving this we would like to observe that Arkin has given a very simple proof of some congruence results concerning the τ-function.

If the C_n are integers and we define C_n^k by the equation

$$\left(\sum_{n=0}^{\infty} C_n x^n\right)^k = \sum_{n=0}^{\infty} C_n^k x^n,$$

then, by differentiating and equating coefficients, one finds, for $k \geq 1$, $n \geq 1$, that

$$C_n^k \equiv 0 \left(\operatorname{mod} \frac{k}{(n,k)}\right).$$

Applying this to the identity defining the τ-function yields

$$\tau(n+1) \equiv 0 \left(\operatorname{mod} \frac{24}{(n,\ 24)}\right).$$

In particular, taking $n = 24t + 8$ we find

$$\tau(3n) \equiv \tau(3n+2) \equiv 0 \pmod{3}.$$

Similarly one finds

$$\tau(2n) \equiv 0 \pmod{8}.$$

See also Integer 691, A Divisibility Property and Integer 63001, Ramanujan's τ function.

REFERENCES

[1] H. P. F. Swinnerton-Dyer, "Congruence properties of $\tau(n)$" in *Ramanujan Revisited* (eds. G. E. Andrews, et al.) Academic Press, Boston, 1988, pp. 289–311.

[2] J. Arkin, Congruences for the coefficients of the kth power of a power series, *Amer. Math. Monthly* 71 (1964) 899–900.

Integer 24

1 Sum of Consecutive Squares a Square

Consider the sequence of numbers

$$1^2,\ 1^2+2^2,\ 1^2+2^2+3^2,\ 1^2+2^2+3^2+4^2,\ 1^2+2^2+3^2+4^2+5^2, \ldots.$$

That is, consider the sequence $1, 5, 14, 30, 55, \ldots$. What can be said about the number of squares in this sequence?

The only value of $n > 1$ for which

$$1^2 + 2^2 + \cdots + n^2$$

is a perfect square is $n = 24$. Thus only the 1st and 24th terms in the sequence are squares.

Though this question was asked by Lucas in 1775, the first correct proof was apparently given by G. N. Watson in 1918. Until 1985 no known proof could be considered elementary. In 1985 De Gang Ma gave such a proof and, in 1990 Anglin gave an even simpler proof. For his proof and a brief history see his paper cited below.

See also Integer 4, A Few Diophantine Equations, and Integer 645, Sum of Consecutive Squares.

REFERENCES

W. Sierpiński, *250 problems in Elementary Number Theory,* American Elsevier, New York, pp. 115.

W. S. Anglin, The square pyramid puzzle, it Amer. Math. Monthly, 97 (1990) 120-4.

2 A Property of 24

The integer 24 is exactly divisible by every integer not exceeding its square root and is the largest positive integer for which this is true.

REFERENCES

J. V. Uspensky and M. A. Heaslet , *Elementary Number Theory*, McGraw-Hill, New York, 1939, Exercise 1, p. 94.

3 Fibonacci Periodicities

If one reduces the terms in the Fibonacci sequence modulo an integer m, one always obtains a periodic sequence. Let $P(m)$ be the period length when one reduces modulo m.

m	2	3	4	5	6	7	8	9	10	11	12	13	14	$\cdots$
$P(m)$	3	8	6	20	24	16	12	24	60	10	24	28	48	$\cdots$.

The sequence

$$m, P(m), P(P(m)), \ldots$$

contains a fixed point and is, therefore, constant from some point on. For $m \geq 2$, $P(m) = m$ if and only if $m = 24 \cdot 5^{\lambda-1}$.

If we reduce each term modulo 7, the resulting sequence starts out as follows:

$$1, 1, 2, 3, 5, 1, 6, 0, 6, 6, 5, 4, \ldots.$$

It will be noted that every possible value from 0 to 6 inclusive appears among these residues. That is, the Fibonacci sequence contains a complete residue system modulo the prime number 7. The same may be said for the primes 2, 3, and 5. However, this statement is not true for any prime greater than 7.

More generally, if one does not confine one's attention to prime moduli, one finds that the only moduli for which the Fibonacci sequence contains a complete system of residues are those of the form $t \cdot 5^k$ where $k \geq 0$ and t is one of 1, 2, 4, 6, 7, 14, or 3^j with $j \geq 1$.

Täcklind has shown that if $a_1, \ldots, a_m, k$ are given integers and the sequence $\{x_n\}$ is defined by

$$x_{n+m} + a_1 x_{n+m-1} + \cdots + a_m x_n = 0$$

and any initial values of the x_j, $1 \leq j \leq n$, then the sequence is periodic modulo k.

In the case $k = 10$, the recurrence $x_{n+2} - x_{n+1} - x_n = 0$ is periodic with period ℓ, where, when not all x_n are even, ℓ is either 60, 12, or 3 depending on whether $x_{n+1} \not\equiv 3x_n \pmod 5$ for all n, $x_{n+1} \equiv 3x_n \not\equiv 0 \pmod 5$, or $x_n \equiv 0 \pmod 5$. In case all x_n are even, the period lengths are 20, 4, and 1 in these respective cases.

From this one sees immediately that the periods of the Fibonacci sequence and of the Lucas sequence $(1, 3, 4, 7, 11, 18, \ldots)$ are 60 and 12, respectively.

If at least one of the first 7 terms is relatively prime to 10, then the sequence with the recurrence $x_{n+7} - x_{n+6} - x_n = 0$ has a period of length greater than 2,480,486.

For more on this topic, see Integer 24, Fibonacci Periodicities, Integer 2935363331541925531, Two Sequences of Complete Integers.

REFERENCES

[1] J. D. Fulton and W. L. Morris, On arithmetic functions related to the Fibonacci numbers, *Acta Arith.* 16 (1969) 105–110.

[2] D. W. Robinson, Iteration of the modular period of a second order linear recurrent sequence, *Acta Arith.* 22 (1973) 249–256.

[3] G. Bruckner, Fibonacci sequence modulo a prime $p \equiv 3 \pmod 4$, *Fib. Quarterly* 8 (1970) 217–220.

[4] S. A. Burr, On moduli for which the Fibonacci sequence contains a complete system of residues, *Fib. Quarterly* 9 (1971) 497–504, 526.

[5] S. Täcklind, Über die Periodizität der Lösungen von Differenzenkongruenzen, *Ark. Mat. Astr. Fys. 30A,* no. 22 (1944).

4 Characters

... for the divisors of 24 and only for them, all characters are real.

—Grosswald

REFERENCES

E. Grosswald, *Topics from the Theory of Numbers* (2nd ed.), Birkhäuser, New York, 1984, p. 248.

5 Divisors of 24

The integer k is a divisor of 24 if and only if, for all ℓ relatively prime to k, it is true that $\ell^2 \equiv 1 \pmod{k}$. (For an equivalent property, see the next subsection.)

Bateman and Low proved Dirichlet's theorem (there exist infinitely many primes in any arithmetic progression for which the first term and the difference are relatively prime) in the special case where the difference is 24. Of this proof, they said:

> For general k a proof of this kind can be given only when
>
> $$\ell^2 \equiv 1 \pmod{k}. \tag{1}$$
>
> Since the only values of k for which (1) holds for *every* ℓ relatively prime to k are $k = 1, 2, 3, 4, 6, 8, 12$, and 24, the modulus 24 is the largest which can be settled completely by the present method.

They further observed that only 24, of all these special values, seems not to have already been individually dealt with in the literature.

Vanden Eynden let $N(n,j,k)$ stand for the number of positive integer divisors d of n for which $d \equiv j \pmod{k}$. He then let S be the number of triples (a,b,k) for which $N(n,a,k) \geq N(n,b,k)$ for all positive n. Proving the above stated property of the integers 1, 2, 3, 4, 6, 8, 12, 24 as a lemma he used it to prove the following theorem.

Theorem.(a,b,k) *is in* S *if and only if* a *is the greatest common divisor of* b *and* k *and* $\frac{k}{a}|24$.

REFERENCES

[1] P. T. Bateman and M. E. Low, Prime numbers in arithmetic progressions with difference 24, *Amer. Math. Monthly* 72 (1965) 139–143.

[2] C. L. vanden Eynden, A congruence property of the divisors of n for every n, *Duke J. Math.* 29 (1962) 199–202.

6 $\sigma(24n-1)$

The totality of integers m for which m divides $\sigma(mn-1)$ for all natural numbers n is $\{3,4,6,8,12,24\}$, i.e., the sum of the divisors of $mn-1$ is divisible by m, where m is any divisor of 24 greater than 2, and by no other m. An immediate consequence is that if m is one of the numbers $3,4,6,8,12,24$, then $m|k+1$ implies that $m|\sigma(k)$. This was (first?) given by Ramanathan [1] in 1943. Ramanathan [2] proved a similar property for the Ramanujan τ function in 1944. I.e., he showed

$$\tau(kn-1) \equiv 0 \pmod{k},$$

where k is any divisor of 24 which is larger than 2. (See Integer 23, The τ Function.) Gupta showed $\sigma(jn-1) \equiv 0 \pmod{j}, j \geq 3$, for all $n > 0$ if and only if $a^2 \equiv 1 \pmod{j}$ for every a less than and prime to j.

As an easy exercise show that when $xy = 24n-1$, all quantities integers, $x+y$ is divisible by 24.

REFERENCES

[1] K. G. Ramanathan, Congruence properties of $\sigma(n)$, the sum of the divisors of n, *Math. Stu.* 11 (1943) 33–35.

[2] ——, Congruence properties of Ramanujan's function $\tau(n)$, *Proc. Ind. Acad. Sci. A* 19 (1944) 146–48.

[3] M. V. Subba Rao, Congruence properties of $\sigma(n)$, *Math. Stu.* 18 (1950/51) 17–18.

[4] H. Gupta, Congruence properties of $\sigma(n)$, *Math. Stu.* 13 (1945) 25–29.

[5] *J. Rec. Math.* 11 (1978) 257, 267 (exercise).

7 Self-dual Lattices in R^n

Suppose Λ is a self-dual lattice in R^n. Then it is always true that

$$\min\{u \cdot u | u \in \Lambda, u \neq 0\} \leq \left[\frac{n}{8}\right] + 1$$

When equality holds the lattice is *extremal.*

The integer 24 is the largest possible dimension for which an extremal self-dual lattice may exist.

REFERENCES

J. H. Conway, A. M. Odlyzko , and N. J. A. Sloane, Extremal self-dual lattices exist only in dimensions 1–8, 12, 14, 15, 23, and 24, *Mathematika* 25 (1978) 36–43.

8 Some Formulae Containing 24

See also Integer 63001, Ramanujan's τ Function.

1. $\pi = 24\tan^{-1}\frac{1}{8} + 8\tan^{-1}\frac{1}{57} + 4\tan^{-1}\frac{1}{239}.$

This formula was used by Maheut in 1976 to compute 500000 digits of *pi* in 24 hours and 10 minutes. It had previously (1962) been used by Shanks and Wrench to find 100000 digits of π.

Newton himself used the formula

$$\pi = \frac{3\sqrt{3}}{4} + 24\left(\frac{1}{12} - \frac{1}{5 \cdot 2^5} - \frac{1}{28 \cdot 2^7} - \frac{1}{72 \cdot 2^9} - \cdots\right)$$

to compute 15 digits of π. He is reported by Borwein, Borwein, and Bailey as saying "I am ashamed to tell you how many figures I carried these calculations, having no other business at the time."

In the last few years much progress has been made on the computation of π to a large number of decimal place and just this morning (October 30, 1989) I received a copy of the MAA newsletter *Focus* reporting on the computation of π by G. V. and D. V. Chudnovsky to over a billion decimal digits (1,011,196,691 to be exact).

The Chudnovskys used the following formula in their calculations.

$$\frac{426880\sqrt{10005}}{\pi} = \sum_{n \geq 0} \frac{(6n!)(545140134n + 13591409)}{n!^3(3n)! \cdot (-640320)^{3n}}$$

and this, in turn, is equal to

$$b - \frac{1}{1}\frac{3}{1}\frac{5}{1}e\left(a + b - \frac{7}{2}\frac{9}{2}\frac{11}{2}e\left(2a + b - \frac{13}{3}\frac{15}{3}\frac{17}{3}e\left(3a + b - \cdots,\right.\right.\right.$$

where $a = 545140$, $b = 13591409$, and $e = 320160^{-3}$. In the *Focus* article (October 1989), it is observed that

> Summing this series to N terms determines π to $14.18N$ decimal places. ... This series arises from the theory of elliptic modular functions and is connected with the quadratic field $Q(\sqrt{-163})$, the largest one-class imaginary quadratic field.

See Integer 100, 100 Digits of π.

2. In a book on the distribution of prime numbers, Huxley proved the following formula involving the number of representations, $d_4(n)$, of the integer n as a product of four factors.

$$\sum_{n \leq x} \frac{d_4(n)}{n} = \frac{1}{24} \log^4 x + O(\log^3 x)$$

3. The partition function $p(n)$ is defined to be the number of ways one may write n as the sum of positive integers. For 5 we have:

$$5 = 4+1 = 3+2 = 3+1+1 = 2+2+1 = 2+1+1+1 = 1+1+1+1+1$$

and, therefore, $p(5) = 7$. This function is very erratic and no simple way of evaluating it is available. Despite considerable interest in the function it was not until about 1917 that anyone questioned the value of the function for large values of n. At that time Hardy and Ramanujan worked out an asymptotic expression for $p(n)$. In 1937 Rademacher gave an exact expression for $p(n)$. The following is a quote from Hardy. (In this quote $\lambda_n = \sqrt{n - \frac{1}{24}}$.)

> Rademacher, who (trying at first merely to simplify our analysis) was led to make a very fortunate formal change. Ramanujan and I worked, not exactly with the function
>
> $$\phi(n) = \frac{1}{2\pi\sqrt{2}} \frac{d}{dn} \left(\frac{e^{K\lambda_n}}{\lambda_n} \right),$$
>
> but with the "nearly equivalent" function
>
> $$\frac{1}{\pi\sqrt{2}} \frac{d}{dn} \left(\frac{\cosh K\lambda_n - 1}{\lambda_n} \right)$$
>
> .... Rademacher works with
>
> $$\frac{1}{\pi\sqrt{2}} \frac{d}{dn} \left(\frac{\sinh K\lambda_n}{\lambda_n} \right),$$
>
> which is also "nearly equivalent"; and this apparently slight change has a very important effect, since it leads to an *identity* for $p(n)$.

In fact, the identity that Rademacher proved is

$$p(n) = \sum_{q=1}^{\infty} \sum \omega_{p,q} e^{-2np\pi i/q} \frac{q^{1/2}}{\pi\sqrt{2}} \frac{d}{dn} \left(\frac{\sinh \frac{K}{q} \sqrt{n - \frac{1}{24}}}{\sqrt{n - \frac{1}{24}}} \right),$$

where the inside sum is over all positive integers q less than and prime to p and $\omega_{p,q}$ is a "certain" 24th root of unity.

REFERENCES

[1] G. Maheut, Quelques considérations sur le nombre π, *Bull. ARERS* 17 (1976) 25–31.

[2] *Focus,* Newsletter of the MAA, 9 (1989) 1–4.

[3] J. M. Borwein, P. B. Borwein, and D. H. Bailey, Ramanujan, modular equations, and approximations to π or how to compute one billion digits of π, *Amer. Math. Monthly* 96 (1989) 201–219.

[4] G. H. Hardy, *Ramanujan: Twelve Lectures on Subjects Suggested by His Life and Works,* Chelsea, New York, (Reprint).

[5] M. N. Huxley, *The distribution of prime numbers,* Oxford Univ. Press, Oxford, 1972, p. 8.

9 A Fermat-like Equation

In an interesting paper, Gandhi discussed the solvability of a number of equations which bear a relationship to Fermat's last theorem (see Integer 1093, Fermat's Conjecture).

As an example of a more general theorem, he proved that the equation

$$x^{24} + y^{24} = pz^{24},$$

where p is any prime, has no nontrivial solutions.

REFERENCES

J. M. Gandhi, On Fermat's last theorem, *Amer. Math. Monthly* 71 (1964) 998–1006.

Integer 25

A Mystery

In 1968 a high school student, Dean Alvis, made the following observation: all primes other than 7 which are less than 2000 have prime base 7 digit sums.

At the same time my son found, on the Reed College computer, that up to 32000 only the prime 4801 had a composite base 7 digit sum. In this case the base 7 digit sum is 25 since the base 7 representation of 4801 is 16666.

Upon careful examination the mystery was resolved as follows.

Let T_n be the base 7 digit sum of n. Then $n - T_n$ is divisible by 6 since it is the sum of multiples of terms of the form $7^t - 1$ (each divisible by 6). For $n > 3$, n prime, neither 2 nor 3 divides n and hence they may not divide T_n. Thus the smallest composite value of T_n must be 5^2 (= 25) for n prime. Hence all primes n with $T_n < 25$ have T_n prime (except 7). Now $T_n < 25$ for $n < 16666$ (base 7), i.e., for $n < 4801$ (base 10).

It is easy to see that 25 is the only possible composite T_n for n prime and smaller than 100841 (base 10).

Integer 27

I Powers as Sums of Digits

The only integers equal to the sum of the digits of their cubes are the numbers

$$1, 8, 17, 18, 26, 27.$$

Mohanty and Kumar state that this was reported by Moret Blanc in 1879. Similar results are given for all powers from 2 to 10. For instance 9^2 has digit sum 9, and each of 22, 25, 28, and 36 is equal to the sum of the digits of their fourth powers. (The topic is also discussed, with incomplete results, by Iseki and Nakakura.)

Each of the two numbers 6859 and 21952 has a digit sum cube equal to the other.

REFERENCES

[1] S. P. Mohanty and H. Kumar, Powers of Sums of Digits, *Math. Mag.* 52 (1979) 310–12.

[2] K. Iseki and M. Nakakura, A simple number theoretic problem, *Math. Japon.* 29 (1984) 835–37.

2 Langford Sequences

In 1958 C. D. Langford posed a problem based on the following observations.

> Years ago, my son, then a little boy, was playing with some coloured blocks. There were two of each colour, and one day I noticed that he had placed them in a single pile so that between the red pair there was one block, two between the blue pair, and three between the yellow. I then found that by a complete rearrangement I could add a green pair with four between them.

He gave the following (numerical) examples.

$n = 3$ 312132
$n = 4$ 41312432
$n = 7$ 17126425374635
$n = 8$ 3181375264285746
$n = 11$ $121e257t8395637e48t694$
$n = 12$ $Tt864e975468tT579e312132$
$n = 15$ $F\theta e975fTt86579e\theta F68tTf41312432$

The general problem, as stated by Hayasaka and Saito [2], is the following.

> Given $2n$ numbers, two each of the numbers $1, 2, \ldots, n$, to find whether they can be arranged in a row in such a way that the two occurrences of each number k are separated by exactly k other elements of the row.

These authors also generalize the problem to a set of sn numbers, s each of $1, 2, \ldots, n$. With the same conditions set on successive occurrences of the same number, such sequences are called *Langford* (s, n)-*sequences.* An example of a Langford (3,9)-sequence is

$$191618257269258476354938743.$$

Prior to 1979, no Langford $(4, n)$-sequences were known to exist. In 1971 it was proved there were neither Langford (4,7)- nor (4,8)-sequences. In 1975 this was extended to the nonexistence of such sequences $(4, n)$ for $n \leq 16$.

In 1979 the 16 in this last result was replaced by 23. It was previously (about 1975) known that there does exist a Langford (4,24)-sequence and in fact that there are exactly three such.

It is known that there are no $(5, n)$-sequences for $n \leq 26$ but that there is a (5,27) sequence.

Given a positive integer n there is an integer $f(n)$ such that there are at least n $(2, f(n))$-sequences.

Various special results have been deduced. One of them is:

> If there is a Langford (s, n)-sequence and p is a prime with $p^e || s$, then n necessarily satisfies one of the congruences
>
> $$n \equiv -1, 0, 1, \ldots, p-2 \pmod{p^{e+1}}.$$
>
> When $n = 2$ the condition is both necessary and sufficient.

A "Langford sequence" with a gap one place from the end is said to be *hooked.* The sequence 1212 would be called a hooked Langford (2,2)-sequence thinking of it as $121 * 2$.

Priday showed that for every n either a Langford $(2, n)$-sequence exists or a hooked Langford $(2, n)$-sequence exists. He further conjectured the first case occurred precisely when $n \equiv 0, -1 \pmod 4$ and the second case precisely when $n \equiv -2, -3 \pmod 4$.

The conjecture was proved by Davies who also noted some connections between the Langford sequences and the so-called Skolem sequences which are defined in the same way except they use the first consecutive nonnegative integers instead of the first consecutive positive integers.

If $V_n = \{v_1, \ldots, v_n\}$, then a string of symbols, each drawn from V_n is called a *weak Langford sequence* if, whenever a symbol v_j occurs more than once, there are exactly j symbols between each pair of successive occurrences of that symbol.

Zsolt has shown that a weak Langford sequence may not contain an 'interior' string consisting of two identical contiguous nonempty blocks of symbols. By 'interior' we mean that neither the first nor the last symbols of the sequence are in the repeated blocks. One calls such a sequence *squarefree.*

Zsolt has also shown that the number of such weak Langford sequences (for fixed n) is finite.

Further generalizations have also been investigated.

REFERENCES

[1] C. D. Langford, Problem, *Math. Gaz.* 42 (1958) 228.

[2] T. Hayasaka and S. Saito, Langford sequences: a progress report, *Math. Gaz.* 63 (1979) 261–62.

[3] ——, Existence of the Langford $(5, n)$-sequence, Res. Rep. Miyagi Tech. Coll. No. 18 (1982) 143–46.

[4] S. Saito, Harmonic analysis of the Langford sequence, Res. Rep. Miyagi Tech. Coll. No. 16 (1980) 99–103.

[5] E. Köhler, Bemerkungen über Langfordsequenzen, Numer. Meth. bei Optimierungsaufgaben, Band 3 (Tagung, Math. Forsch., Oberwohlfach, 1976) 137–145.

[6] S. Saito and T. Hayasaka, The Langford $(4, n)$-sequence: a trigonometric approach, *Disc. Math.* 28 (1979) 81–88.

[7] C. J. Priday, On Langford's problem I, *Math. Gaz.* 43 (1959) 250–53.

[8] R. O. Davies, On Langford's problem II, *Math. Gaz.* 43 (1979) 253–55.

[9] T. Zsolt, Langford strings are squarefree, *Int. J. Comp. Math.* 29 (1989) 75–78.

Integer 28

A Special Perfect Number

The integer 28 is the only even perfect number of the form $x^n + y^n$, where n is greater than one and x and y are relatively prime. If one puts

$$\frac{\sigma(N) - N}{N} = h,$$

then, when h is an integer, one calls N *h-fold perfect*. In this terminology a perfect number is 1-fold perfect. When the canonical prime factorization of N contains n distinct prime factors one says that N has *rank* n. A theorem of Dickson and Gradstein tells us that there are at most a finite number of perfect numbers among those of given rank. This has been extended by Artuhov, see citation below, who proved the following theorem.

Theorem. 1. *For given natural numbers n, h there are at most finitely many h-fold perfect numbers of rank n.*

2. *For given natural number n there are only finitely many h-fold perfect numbers of rank n with odd h.*

Artuhov's article gives references to tables of h-fold perfect numbers for h not exceeding 7.

Another generalization of perfect number is as follows. The integer m is an *n-hyperperfect number* if

$$m = 1 + n(\sigma(n) - m - 1).$$

In this terminology the perfect numbers are the 1-hyperperfect numbers.

At one time there was a conjecture that all such numbers contained no more than two prime factors. This was disproved by te Riele.

See also Integer 6, Perfect Numbers and Integer 90, Bi-unitary Perfect Numbers.

REFERENCES

[1] T. N. Sinha, Note on perfect numbers, *Math. Stu.* 42 (1975) 336.

[2] M. M. Artuhov, On the problem of h-fold perfect numbers, *Acta Arith.* 23 (1973) 249–255.

[3] H. J. J. te Riele, Hyperperfect numbers with three different prime factors, *Math. Comp.* 36 (1981) 297–98.

Integer 29

Consecutive Integers

If one examines the first 29 consecutive positive integers, one finds that no one of them has more than two different prime divisors. Indeed, the smallest integer with three different prime divisors is $2 \cdot 3 \cdot 5$.

Sierpiński observed that there does not exist a longer string of consecutive integers with this property.

Since, for all n, n divides exactly one of any string of n consecutive integers the following is true. If $m(q)$ is the longest string of consecutive integers such that no integer in the string has more than q distinct prime divisors, then $m(q) + 1$ is the product of the first $q + 1$ prime numbers and the string $1, 2, 3, \ldots, m(q)$ is the only string having the property.

Integer 30

1 Property of 30

The integer 30 is the largest integer for which all smaller integers ($\neq 1$) relatively prime to it are themselves prime. This was first proved by Schatunowsky in 1893.

A very simple proof of this, using an elementary inequality due to Bonse, may be found in the book by Uspensky and Heaslet [2].

This result has been generalized. Landau gave a proof of the following theorem which was proved by Maillet in 1900 (see Dickson [1, p. 134]).

> For fixed r there are only finitely many positive integers n for which all of the $\phi(n)$ positive integers not exceeding n and relatively prime to n have at most r prime factors (repetitions counted).

The case $r = 1$ yields only the integers $1, 2, 3, 4, 6, 8, 12, 18, 24, 30$. Recently Iwata discussed this generalization of Landau.

See also Integer 18, Totitives.

REFERENCES

[1] L. E. Dickson, *History of the Theory of Numbers,* vol. I, Chelsea, New York, 1952, p. 132 (original edition 1918).

[2] J. V. Uspensky and M. A. Heaslet , *Elementary Number Theory,* McGraw-Hill, New York, 1939.

[3] E. Landau, *Handbuch der Lehre von der Verteilung der Primzahlen,* vol. I, Chelsea, New York, 1953, pp. 229–234 (original edition 1909).

[4] H. Iwata, On Bonse's theorem, *Math. Rep. Toyama Univ.* 7 (1984) 115–117.

2 A Spurious Property

As an outgrowth of a question asked by Sierpiński a computer search by Lander and Parkin in 1967 found the smallest arithmetic progression of length 6 consisting solely of primes to be $121174811 + 30t$, $0 \leq t \leq 5$. Altogether they found 27 such progressions and they all had difference equal to 30. The least arithmetic progression of primes and of length 5 is $9843019 + 30t$, $0 \leq t \leq 4$. (It is not difficult to show that if 5 **consecutive** primes are in arithmetic progression, then the difference for the progression must be divisible by 30. The reader might try this as an exercise. Mąkowski gave the following example of four consecutive primes in arithmetic progression: n, $n + 30$, $n + 60$, $n + 90$, where $n = 6248969$.)

In November 1990 A. Moran and P. Pritchard found the following sequence of 21 primes (the old record had 19). (See Proceedings of the 14th Australian Computer Conference 1991, to appear.)

$$142072321123 + 1419763024680t, \qquad 0 \leq t \leq 20.$$

In his book, Sierpiński showed that if there is such a progression of length 100, then the difference would have to have more than 30 digits.

We let $N_m(x)$ be the number of m termed arithmetic progressions consisting only of primes which do not exceed x. In 1938 Estermann showed $N_3(x)$ is of order greater than $x^2/(\log x)^3$ and this estimate was improved in 1982 by Grosswald . Thus, there are infinitely many three term arithmetic progressions of primes. It is not known if there are infinitely many four term arithmetic progressions of primes. However, Heath-Brown showed, in 1981, that there are infinitely many four term arithmetic progressions consisting of three primes and a number which is either a prime, a square of a prime, or a product of two primes. The possible non-prime is either the first or last of the four numbers.

REFERENCES

[1] A. Mąkowski, *Math. Gaz.* 44 (1960) 220.

[2] P. A. Pritchard, Long arithmetic progressions of primes: some old, some new, *Math. Comp.* 45 (1985) 263–67.

[3] E. Grosswald, Arithmetic progressions that consist only of primes, *J. Number Thy.* 14 (1982) 9–31.

[4] W. Sierpiński, *Elementary Theory of Numbers,* North-Holland, Amsterdam, 1988, pp. 126–28.

[5] D. R. Heath-Brown, Three primes and an almost prime in arithmetic progression, *J. London Math. Soc.* (2) 23 (1981) 396–414.

Integer 32

A Conjecture of Carmichael

In 1907, in the *Bulletin of the American Mathematical Society* (vol. 13, pp. 241–43), R. D. Carmichael conjectured that the equation $\phi(x) = n$ never has exactly one solution for x. Grosswald has shown that if there is an n for which this equation has exactly one solution, then 32 must be a divisor of n.

REFERENCES

E. Grosswald, Contribution to the theory of Euler's function $\phi(x)$, *Bull. Amer. Math. Soc.* 79 (1973) 337–341.

Integer 33

Complete Sequences

For $S = \{s_1, s_2, \ldots\}$, a sequence of real numbers, let $P(S)$ be the set of all numbers representable as a sum of a finite number of distinct elements of S. The set S is *complete* if all sufficiently large integers are in $P(S)$. For f a real polynomial, we write $S(f) = \{f(1), f(2), \ldots\}$.

When S is an infinite set of positive integers, Richert proved the following theorem in 1949.

Theorem. *Suppose there are integers k, N, M with $k \geq 2$, $N \geq 0$, $M > 0$ and such that the following three properties are satisfied.*

1. *If $N < x \leq N + M$, then x is a sum of distinct elements chosen from among the first k elements of S.*
2. $M \geq s_{k+1}$.
3. $2s_i \geq s_{i+1}$ *for* $i > k$.

Then every integer greater than N is a sum of distinct elements of S.

When $S = \{1, 3, 6, 10, 15, 21, 28, 36, 45, \ldots\}$, where the nth term of the sequence is the nth triangular number, we may choose $k = 8$, $N = 33$, $M = 45$ to see that all integers greater than 33 are writable as sums of distinct triangular numbers. Hence the set of triangular numbers is complete; the only positive integers not sums of triangular numbers being 2 , 5, 8, 12, 23, 33.

Even easier, we may show that the Fibonacci sequence is complete by taking $k = 3$, $N = 0$, $M = 4$. In fact, by a result of Zeckendorf every natural number has a unique representation as a sum of nonconsecutive (i.e., no two are consecutive) Fibonacci numbers. Daykin has shown that the Fibonacci sequence is the only sequence with this property.

Using Richert's theorem, Dressler and Parker showed 12758 to be the largest integer not representable as a sum of distinct cubes. (This had been shown some ten years earlier by Graham.)

Sprague had proved in 1949 that for each $k \geq 2$ there is a largest integer, say $r(n)$, which is not a sum of distinct kth powers. The preceding statement is just $r(3) = 12758$.

Brown has proved the following generalization of the above.

Let $S_1, S_2, \ldots$ be a sequence of positive integers and assume there are fixed integers $r \geq 0, K_0 \geq 0$ such that the following two properties are satisfied.

1. $S_{n+1} \leq S_{r+1} + \sum_{r+1}^{n} S_i$ for all $n \geq r$.

2. Each integer N in the range $K_0 \leq N < K_0 + S_{r+1}$ has an expansion in the form $N = \sum_1^r \alpha_i S_i$.

Then any positive integer N satisfying $K_0 \leq N < K_0 + S_{r+1} + \sum_{r+1}^{n} S_i$ for some $n \geq r$ is expressed in the form $N = \sum_1^n \alpha_i S_i$. (All α_i are either 0 or 1.)

Using this Brown has proved a number of results such as those in Integer 6, Sums of Distinct Primes and Integer 128, Sums of Distinct Squares. In the notation introduced above $r(2) = 128$.

Graham has proved the following results about complete sequences.

Theorem 1. *If $f(x) = \alpha_n x^n + \cdots + \alpha_1 x + \alpha_0, \alpha_n \neq 0$, and f takes integers into integers, then $S(f)$ is complete if and only if $\alpha_n > 0$ and for every prime p there is an m such that p does not divide $f(m)$.*

Theorem 2. *Let*

$$f(x) = \frac{p_0}{q_0} + \frac{p_1}{q_1}\binom{x}{1} + \cdots + \frac{p_n}{q_n}\binom{x}{n},$$

where for each j, the quantities p_j, q_j are relatively prime integers, $p_n \neq 0$, and $q_k \neq 0$ for all k. Then $S(f)$ is complete if and only if $\frac{p_n}{q_n} > 0$ and the gcd *of all the p_j is 1.*

Theorem 3. *If f is as in Theorem 2 and at least one of the α_k is irrational, then $S(f)$ is not complete.*

Theorem 4. *If $f(x) = \alpha_0 + \alpha_1\binom{x}{1} + \cdots + \alpha_n\binom{x}{n}$, $\alpha_n \neq 0$, all α_k real, then $S(f)$ is complete if and only if all of the following are the case.*

1. *For $0 \leq k \leq n$ there are relatively prime integers p_k, q_k, with the $q_k \neq 0$, such that $\alpha_k = \frac{p_k}{q_k}$.*
2. $\alpha_n > 0$.
3. $\gcd(p_0, p_1, \ldots, p_n) = 1$.

Corollary. *$\{f(1), f(2), \ldots\}$ is complete if and only if for every n, $\{f(n), f(n+1), \ldots\}$ is complete.*

Let $\lambda(f)$ be the largest integer not in $P(S(f))$. Then the following values of this quantity are known.

$$\lambda(\frac{x^2+x}{2}) = 33 \qquad \text{(see Richert's result above);}$$

$$\lambda(x^2) = 128 \qquad \text{(Sprague 1948);}$$

$$\lambda(x^3) = 12758 \qquad \text{(Graham);}$$

$$\lambda(x^4) = 5134240 \qquad \text{(Lin 1970);}$$

$$\lambda(ax - a + 1) = \frac{a^2(a-1)}{2}.$$

Thus the largest integer not the sum of distinct triangular numbers is 33; not the sum of distinct squares is 128; and not the sum of distinct 4th powers is greater than five million.

The polygonal numbers of order $k > 2$ are those of the form

$$x + \frac{(k-2)(x^2 - x)}{2}.$$

Letting the largest integer not the sum of distinct polygonal numbers of order k be denoted by s_k Mimura gives the table

k	3	4	5	6	7	8	9	10	11	12	13	14	15
s_k	33	128	159	267	387	713	1152	929	994	1240	1770	1943	1950

Brown and Weiss call a sequence of positive integers an *N-sequence* if after any removal of N terms, and after no removal of $N + 1$ terms, all positive integers are sums of the terms remaining. They then show that, though the Fibonacci sequence is known to be a 1-sequence, N-sequences, for $N > 1$, do not exist.

REFERENCES

[1] R. L. Graham, Complete sequences of polynomial values, *Duke Math. J.* 31 (1964) 175–285.

[2] H. E. Richert, Über Zerlegungen in paarweise verschiedene Zahlen, *Norsk Mat. Tidsskr.* 31 (1949) 120–22.

[3] D. E. Daykin, Representation of natural numbers as sums of generalized Fibonacci numbers, *J. London Math. Soc.* 35 (1960) 143–160.

[4] R. E. Dressler and T. Parker , 12758, *Math. Comp.* 28 (1974) 313–314.

[5] R. Sprague, Über Zerlegungen in ungleiche Quadratzahlen, *Math. Zeit.* 51 (1948) 289–290, 466–468.

[6] J. L. Brown, Generalization of Richert's theorem, *Amer. Math. Monthly* 83 (1976) 631–634.

[7] S. Lin, Computer experiments on sequences which form integral bases, in *Computer Problems in Abstract Algebra* (ed. J. Leech), Pergamon, Oxford, 1970, pp. 365–370.

[8] T. Kløve, Sums of distinct elements from a fixed set, *Math. Comp.* 29 (1975) 1144–1149.

[9] Y. Mimura, Partial sums of sequences, *Math. Japonica* 36 (1991) 677–679.

[10] T. C. Brown and M. L. Weiss, On N-sequences, *Math. Mag.* 44 (1971) 89–92.

Integer 36

Sums of Digits

Let $S_p(n)$ be the sum of the base p digits of n. If $S_p(n) < p$ for all odd primes p, or even just for the odd primes $3, 5, 7, 11, 13$, then n must be one of the 10 integers

$$1, 2, 3, 4, 6, 10, 12, 28, 30, 36.$$

REFERENCES

J. J. Schäffer, A result in elementary number theory, *Nieuw Arch. Wisk.* 4 (1956) 118–123.

Integer 38

Sum of Composites

Every even integer is of one of the forms

$$10k,\ 10k+2,\ 10k+4,\ 10k+6,\ 10k+8.$$

Noting that

$$\begin{aligned} 10k &= 15 + (10k - 15) \\ 10k + 2 &= 27 + (10k - 25) \\ 10k + 4 &= 9 + (10k - 5) \\ 10k + 6 &= 21 + (10k - 15) \\ 10k + 8 &= 33 + (10k - 25) \end{aligned}$$

we see that every even integer greater than 38 may be written as the sum of two composite odd integers.

In fact, exactly 8 positive even integers are not writable as the sum of two composite odd integers, and the largest of these 8 is 38. The set of 8 integers is

$$\{4, 6, 8, 12, 14, 20, 32, 38\}.$$

More generally, Vaidya [2] proved the following.

1. Among all integers having the same parity as t ($t \geq 2$, t an integer) the largest not expressible as a sum of t odd composite integers is $9t + 20$.

2. If k is a positive integer and p, q are odd primes, then the largest even integer that has fewer than k representations as a sum of two odd composite numbers, one a multiple of p and the other a multiple of q, is $2kpq + q + p$.

Later, Vaidya [3] showed that every sufficiently large positive integer may be written as a sum of two relatively prime composite integers.

REFERENCES

[1] E. Just and N. Schaumberger, A curious property of the integer 38, *Math. Mag.* 46 (1973) 221.

[2] A. M. Vaidya, On representing integers as sums of odd composite integers, *Math. Mag.* 48 (1975) 221–23.

[3] ——, "Representing integers as sums of two relatively prime composite integers" in *Proc. of the Second Conference on Number Theory,* R. Jagannathan (ed.), Ootacamund, 1980, pp. 119–121.

Integer 47

Cutting a Cube into Cubes

Not insisting that the subcubes be different in size, Eves said the following about cutting a cube into cubes.

> It is easy to cut a cube into 8 subcubes (by planes parallel to pairs of opposite faces and midway between them), and it is quite apparent that it is impossible to cut a cube into only 2 subcubes. For a given positive integer k, then, it may or may not be possible to cut a cube into k subcubes. William Scott, in 1946, showed that a cube can be cut into any $k > 64$ subcubes. It was shown also that a cube can be cut into k subcubes where k is any of the following integers.
>
> $$1, 8, 15, 20, 22, 27, 29, 34, 36, 38, 39$$
> $$41, 43, 45, 46, 48, 49, 50, 51, 52, 53.$$
>
> It was natural, then, to wonder if $k = 54$ is the largest number of subcubes into which a cube cannot be cut. This question was still unanswered when the first edition of the present book appeared in 1968. Reading this and challenged by the question, Von Christian Thiel, of Germany, attacked the problem and in

1969 managed to show that a cube can be cut into 54 subcubes. The question now is: Is $k = 47$ the largest number of subcubes into which a given cube cannot be cut?

See also Integers 2,3, Cubing Cubes and Integer 23, Filling Boxes.

REFERENCES

H. Eves, *A Survey of Geometry,* (rev. ed.), Allyn and Bacon, Boston, 1972, pp. 236–37.

Integer 52

A Linear Recurrence

Mignotte studied the recurrence

$$u_0 = u_1 = 0, \qquad u_2 = 1, \qquad u_{n+3} = 2u_{n+2} - 4u_{n+1} + 4u_n.$$

This is the so-called *recurrence of Berstel.* Previously it was known that $u_n = 0$ precisely for $n = 0, 1, 4, 6, 13, 52$.

Mignotte determined for which $c\, u_n = \pm c$ has at least two solutions.

Van der Poorten's paper [2, p. 516] gives a little further information on this recurrence.

REFERENCES

[1] M. Mignotte, Determination des répetitions d'une certaine suite recurrente linéaire, Publ. Math. Debrecen 33 (1986) 297–306.

[2] A. J. van der Poorten, "Some facts that should be better known, especially about rational functions" in *Number Theory and Applications* (ed. R. A. Mollin), Kluwer, Dordrecht, 1988, pp. 497–528.

Integer 53

3-Solubility

Consider a splitting of the first n positive integers into three disjoint sets. E. G. Straus showed that if $n = 54$, then the equation $x+y = 3z$ may be solved in such a way that all of the quantities x, y, z are taken from the same one of the three sets. One says that this equation is *3-soluble* in the set of the first 54 positive integers. He also showed that this equation is not 3-soluble in the set of the first 53 positive integers.

Let A_1, A_2, A_3 be given by

$$A_1 = \{x|x \equiv 1 \pmod 3\} \cup \{x|x \equiv 3 \pmod 9\}$$

$$A_2 = \{x|x \equiv 2 \pmod 3\} \cup \{9, 27, 36\}$$

$$A_3 = \{6, 15, 18, 24, 33, 42, 45, 51\}.$$

With the induced splitting of the set of the first 53 positive integers, the equation $x + y = 3z$ is not solvable with x, y, z all in the same one of these three sets. I.e., the equation is not 3-soluble in the set of the first 53 positive integers. (See Rado below.)

A set S is called *sum-free* if for all pairs a, b in S it is always the case that $a + b$ is not in S.

The first 4, but not the first 5, consecutive integers may be split into 2 sum-free sets.

The first 13, but not the first 14, consecutive integers may be split into 3 sum-free sets.

The first 44, but not the first 45, consecutive integers may be split into 4 sum-free sets.

In the last instance the splitting may be done in exactly 273 different ways whereas in the second in only three ways and in the first the splitting is unique.

Bačik called a set A of positive integers *strongly sum-free* if no element of A is representable as a sum of two *distinct* elements of A. Given k let $f(n, k)$ be the length of the longest interval $[n, f(n, k)]$ which is decomposable into k strongly sum-free sets. He showed that $f(1, 3) = 23$ and $f(n, 3) = 14n + 7$ and conjectured that $f(1, 4) = 66$, $f(n, 4) = 41n + 21$, etc.

REFERENCES

[1] R. Rado, Some partition theorems, in *Combinatorial Theory and its Applications* III (eds. P. Erdős, A. Rényi, and V. Sós), North-Holland, Amsterdam, 1970, pp. 929–936.

[2] H. Fredrickson, Five sum-free sets of $\{1, \ldots, 157\}$, *J. Comb. Thy. A* 27 (1979) 376–77.

[3] W. D. Wallis, A. P. Street , and J. S. Wallis, *Combinatorics: Room Squares, Sum-free Sets, Hadamard Matrices,* LNM #292, Springer, New York, 1972.

[4] J. Bačik, The decomposition of an interval into three strongly sum-free sets, *Acta Math. Univ. Comenian.* 48/49 (1986) 23–35 and (1987).

Integer 67

An Inequality on $\pi(x)$

It is not known whether or not

$$\pi(x + y) \leq \pi(x) + \pi(y)$$

for all sufficiently large x, y. Extending work of Rosser and Schoenfeld, various authors have proved related results. Krawczyk proved the following,

$$\pi(kx) < k\pi(x) \qquad \text{for } k \geq e, x \geq 67.$$

For $\epsilon > 0$, $x \geq 17$, $y \geq 17$, $x + y \geq 1 + e^{4(1+1/\epsilon)}$, it has been shown that

$$\pi(x + y) < (1 + \epsilon)(\pi(x) + \pi(y)).$$

There is an interesting connection between Conjecture A and the generalization of the twin prime conjecture that follows it.

Conjecture A. $\pi(x + y) \leq \pi(x) + \pi(y)$, $x, y \geq 2$.

Conjecture B (Twin Prime Conjecture). *There are infinitely many positive integer values of n such that*

$$n + b_1, \ldots, n + b_k$$

are all prime if and only if for every prime number p there is a congruence class modulo p which contains none of the b_i.

Hensley and Richards [3] showed that these two conjectures are incompatible. The authors said, "We lean towards the opinion that the k-tuples conjecture is true, and (A) is false." An informal, and very readable, account of how this result was obtained is given in the first reference below. The last sentence of Hensley and Richards' article is: "We could have solved the problem without the computer, but we probably wouldn't have."

REFERENCES

[1] I. Richards, On the incompatibility of two conjectures concerning primes; a discussion of the use of computers in attacking a theoretical problem, *Bull. Amer. Math. Soc.* 80 (1974) 419–438.

[2] D. Hensley and I. Richards, On the incompatibility of two conjectures, Proc. Symp. in Pure Math. xxiv, Amer. Math. Soc. 1973, pp. 123–27.

[3] ——, Primes in intervals, *Acta Arith.* 25 (1973/74) 375–391.

[4] A. Krawczyk, On some properties of the function $\pi(x)$, Pr. Nauk. Inst. Math., Politech. Wroclaw 12 Ser. Stud. Mater. 11 (1976) 49–50.

[5] V. Şt. Udrescu, Some remarks concerning the conjecture $\pi(x + y) \leq \pi(x) + \pi(y)$, *Rev. Roumaine Math. Pures. Appl.* 20 (1975) 1201–1209.

Integer 70

A Curious Property

Put $a_0 = n$ and suppose $a_0, \ldots, a_{k-1}$ are specified. Define a_k as the least integer larger than a_{k-1} which is relatively prime to $a_0 \cdots a_{k-1}$.

Theorem. *The only integers n for which every a_k, $k \geq 1$, is either a prime or a power of a prime are*

$$3, 4, 6, 7, 8, 12, 15, 18, 22, 24, 30, 70.$$

These sequences are examined in greater detail in a paper by Erdős, Penney, and Pomerance.

REFERENCES

[1] P. Erdős, A property of 70, *Math. Mag.* 51 (1978) 238–240.

[2] P. Erdős, D. E. Penney, and C. Pomerance, On a class of relatively prime sequences, *J. Number Theory* 10 (1978) 451–474.

Integer 77

1 Unit Fractions

Some integers may be partitioned into a sum of distinct integers whose reciprocals total 1. For example:

$$11 = 2 + 3 + 6, \qquad 1 = \frac{1}{2} + \frac{1}{3} + \frac{1}{6}$$

and

$$24 = 2 + 4 + 6 + 12, \qquad 1 = \frac{1}{2} + \frac{1}{4} + \frac{1}{6} + \frac{1}{12}.$$

Graham [1] showed that every integer greater than 77 may be partitioned into distinct positive integers whose reciprocals add to 1. However, Lehmer carried out some calculations to show that 77 itself does not have such a partition.

More generally, Graham showed: if α, β are positive rational numbers, then there exists a number r, depending on α and β, such that every number greater than r may be partitioned into integers larger than β whose reciprocals add to α.

Burshtein showed the existence of 79 positive integers, no one of them dividing another, with reciprocal sum equal to 1. (This was in answer to a question of Erdős.)

In the same paper, he observed that 945, the smallest odd abundant integer, has the set $\{ 3, 5, 7, 9, 15, 21, 27, 35, 63, 105, 135 \}$ of divisors whose reciprocals sum to 1.

Later, Graham [2] discussed the representation of numbers by finite sums of reciprocals of nth powers of distinct integers. A particularly interesting special case is when $n = 2$. Here he proved that a rational number r is representable as a finite sum of reciprocals of distinct integer squares if and only if it belongs to the union of the two intervals $[0, \frac{\pi^2}{6} - 1]$ and $[1, \frac{\pi^2}{6}]$.

Sierpiński showed that the number of representations of those rationals is infinite. He also showed that every rational between 0 and 2, with the exception of 1, has infinitely many representations as a finite sum of reciprocals of distinct triangular numbers.

The reciprocal of a positive integer is called a *unit fraction.* That every positive integer is a finite sum of distinct unit fractions has been known since 1202. As a model of good exposition we would like to give a longish quote from a paper by Botts in which he lays out the problem and his method of attacking it.

> If we ... set $n = 1$ in the identity
>
> $$\frac{1}{n} = \frac{1}{n+1} + \frac{1}{n(n+1)} \tag{1.1}$$
>
> we have $1 = \frac{1}{2} + \frac{1}{2}$. This expresses 1 as a sum of unit fractions, i.e., fractions with numerator 1; but they are not distinct unit fractions. Using the same identity with $n = 2$ to replace one of these, we get $1 = \frac{1}{2} + \frac{1}{3} + \frac{1}{6}$, which now expresses 1 as a sum of *distinct* unit fractions, the smallest of their denominators being 2. We may then use this last expression and a succession of substitutions from (1.1) to express 1 as a sum of distinct unit fractions in which the smallest denominator is 3, as follows:
>
> $$1 = \frac{1}{2} + \frac{1}{3} + \frac{1}{6} = \left(\frac{1}{3} + \frac{1}{6}\right) + \left(\frac{1}{4} + \frac{1}{12}\right) + \left(\frac{1}{7} + \frac{1}{42}\right)$$
> $$= \frac{1}{3} + \frac{1}{4} + \frac{1}{6} + \frac{1}{7} + \frac{1}{12} + \frac{1}{42}.$$
>
> The reader (with enough paper) may similarly obtain expressions for 1 as a sum of distinct unit fractions in which the

smallest denominator is, successively, $4, 5, 6, 7$. The question arises: can this process be continued indefinitely, so as to yield for each positive integer n an expression for 1 as a finite sum of distinct unit fractions, the smallest denominator being n? Or does the process at some point "blow up," leading to an endless succession of substitutions from (1.1)?

We shall show ... that this process can indeed be continued indefinitely ... Once we have established that the integer 1 is expressible as a finite sum of distinct unit fractions with arbitrarily large denominators, we see at once that there is a similar expression for any positive integer p: we simply think of p as the sum of p 1's. Having found such an expression for an arbitrary positive integer p, we may divide through by an arbitrary positive integer q and in this way obtain a similar expression for any positive rational number whatever. We thus arrive at the following theorem.

Theorem. *For any positive rational number r and any positive integer n, there is an expression for r as a finite sum of distinct unit fractions, all with denominators greater than n.*

This is, in strengthened form, a result first obtained by Fibonacci ... in the year 1202 and essentially rediscovered by Sylvester ... in 1880.

Van Albada and van Lint showed that every integer can be written as the sum of reciprocals of finitely many integers taken from an arbitrary arithmetic progression. They showed that the number of summands need not exceed $e^{n-\gamma}(1 + O(\frac{1}{n}))$, where γ is Euler's constant.

There is a very large literature on the subject.

The following "Egyptian algorithm for polynomials" was given by Dobbs and McConnel. In the statement R is an integral domain with the quotient field K.

Theorem 1.1. *Each element of $K(X)$ can be expressed as a sum of the form*

$$\frac{h_0}{r_0} + \frac{r_1}{h_1} + \cdots + \frac{r_m}{h_m}$$

where $r_i \in R$ for each i and $r_0 \neq 0$, $h_i \in R[X] \setminus \{0\}$ for each $i \geq 1$ and $h_i \in R[X]$, and $deg(h_1) < deg(h_2) < \cdots < deg(h_m)$.

For proper fractions $h_0 = 0$.
Representations are not unique, witness

$$\frac{X^2 + X + 1}{X^3} = \frac{1}{X} + \frac{1}{X^2} + \frac{1}{X^3} = \frac{1}{X - 1} + \frac{1}{X^3(-X + 1)}.$$

See Integer 20161, Abundant Numbers.

REFERENCES

[1] R. L. Graham, A theorem on partitions, *J. Aust. Math. Soc.* 3 (1963) 435–441.

[2] N. Burshtein, On distinct unit fractions whose sum is 1, *Disc. Math.* 5 (1973) 201–206.

[3] R. L. Graham, On finite sums of reciprocals of distinct nth powers, *Pac. J. Math.* 14 (1964) 85–92.

[4] W. Sierpiński, Remarques sur une problème de M. P. Erdős, *Publ. Inst. Math.* (Beograd) (N.S.) 4 (18) (1964) 125–134.

[5] T. Botts, A chain reaction process in number theory, *Math. Mag.* 40 (1967) 55–65.

[6] P. J. van Albada and J. H. van Lint, Reciprocal bases for the integers, *Amer. Math. Monthly* 70 (1963) 170–74.

[7] D. E. Dobbs and R. M. McConnel, An Egyptian algorithm for polynomials, *Elem. Math.* 39 (1984) 126–29.

2 Sums of Squares

Exactly 77 positive integers are not equal to the sum of 5 unequal squares. The largest of these integers is 224.

REFERENCES

G. Pall, On sums of squares, *Amer. Math. Monthly* 40 (1933) 10–18.

Integer 84

Automorphisms of a Curve

In 1893 Hurwitz showed that an algebraic curve over the complex numbers of genus $g \geq 2$ cannot have more than $84(g-1)$ automorphisms. (Here, *automorphism* means birational self-transformation.)

The bound is not attainable for all g. In 1879 Klein showed that $x^3y + y^3z + z^3x = 0$ was of genus 3 and had 168 ($= 84 \cdot 2$) automorphisms. Wiman showed that among the integers from 2 to 6 only for 3 is the bound attainable. Macbeath gives an example of a curve of genus 7 with 504 ($= 84 \cdot 6$) automorphisms. The group of automorphisms is isomorphic to $LF(2, 2^3)$.

The Hurwitz theorem has been extended to fields other than the complex numbers by Roquette.

REFERENCES

[1] A. Hurwitz, Über algebraische Gebilde mit eindeutigen Transformationen in sich, *Math. Ann.* 4 (1893) 403–442.

[2] A. M. Macbeath, On curves of genus 7, *Proc. London Math. Soc.* 15 (1965) 527–542.

[3] P. Roquette, Abschätzung der Automorphismenanzahl von Funktionenkörpern bei Primzahlcharakteristic, *Math. Zeit.* 117 (1970) 157–163.

Integer 86

A School Girl

R. K. Guy related the following story.

> There used to be an admission examination, the "11-plus," to British secondary schools. A question that was asked on one occasion was "Take 7 from 93 as many times as you can." One child answered, "I get 86 every time." I hope she got her place!

REFERENCES

R. K. Guy, Conway's prime producing machine, *Math. Mag.* 56 (1983) 26–33.

Integer 87

Prime Proof

Jones, Sato, Wada, and Wiens showed that if p is a prime number, then there is a proof of its primality consisting of only 87 additions and multiplications.

In 1971 Julia Robinson, commenting on the 1970 negative proof of Hilbert's tenth problem by Matijasevič, observed, "If m is composite, there is a proof that m is composite consisting of one multiplication. Until now it was not known if there was a similar proof consisting of a bounded number of additions and multiplications to show that a prime p is prime."

Jones, Sato, Wada, and Wiens give a polynomial of degree 25 in 26 variables whose positive range is exactly the set of prime numbers. One can reduce the degree of such a polynomial to 5 by using a cute trick going back to Skolem. Applied to the particular polynomial mentioned above this raises the number of variables to 42. The minimum number of variables needed is not known nor is it known if there exists such a polynomial of degree smaller than 5. Matijasevič has shown, however, that no more than 10 variables are needed.

It should be emphasized that polynomials such as we have been speaking about will take other values than just the primes. The above mentioned polynomial, for example, takes the value -76.

In a similar vein, polynomials have been given whose positive range consists, for example, precisely of: the Mersenne primes; the Fermat primes; the even perfect numbers; the Fibonacci numbers; the Lucas numbers; and the perfect numbers.

As a particularly simple example, the polynomial

$$2xy^4 + x^2y^3 - 2x^3y^2 - y^5 - x^4y + 2y \qquad x \geq 1, y \geq 1$$

has the Fibonacci numbers as positive range.

See also Integer 4, Diophantine Representation; Integer 5, Primes as Range of a Polynomial; Integer 23, Hilbert's List; and Integer 100, Recursive Sets.

REFERENCES

[1] J. P. Jones, D. Sato, H. Wada, and D. Wiens, Diophantine representation of the set of prime numbers, *Amer. Math. Monthly* 83 (1976) 449–464.

[2] J. Robinson, "Hilbert's tenth problem" in *Proc. Symp. Pure Math.* vol. 20, Amer. Math. Soc., Providence, 1971, pp. 191–94.

[3] J. V. Matijasevič, Primes are enumerated by a polynomial in 10 variables, *Zap. Naučn. Sem. Leningrad. Otdel. Mat. Inst. Steklov* (LOMI) 68 (1977) 62–82, 144–45.

[4] J. P. Jones, Diophantine representation of Mersenne and Fermat primes, *Acta Arith.* 35 (1979) 209–221.

[5] ——, Diophantine representation of the Fibonacci (Lucas) numbers, *Fib. Quarterly* 13 (1975) 84–88 and *Fib. Quarterly* 14 (1976) 134.

[6] V. Ju. Krjaučjukas, A Diophantine representation of perfect numbers (in Russian), *Zap. Naučn. Sem. Leningrad. Otdel. Mat. Inst. Steklov.* (LOMI) 88 (1979) 78–89, 239.

Integer 88

Quadratic Forms

Exactly 88 forms, $ax^2 + by^2 + cz^2 + dw^2$, represent all but one integer.

For instance, $x^2 + 2y^2 + 7z^2 + 13w^2$ represents every positive integer with the exception of 5.

See also Integer 5, A Conjecture based on Goldbach's Conjecture; Integers 15, 54, and 88, Universal and Almost Universal Forms; and Integer 23, Birthday Problems.

REFERENCES

P. R. Halmos, Note on universal forms, *Bull. Amer. Math. Soc.* 44 (1938) 141–44.

Integer 90

Bi-unitary Perfect Numbers

If d is a divisor of n and the numbers d and $\frac{n}{d}$ have greatest common divisor 1, we call d a *unitary divisor* of n. If the greatest common unitary divisor of d and $\frac{n}{d}$ is 1, we call d a *bi-unitary divisor* of n.

In analogy with the definition of $\sigma(n)$ as the sum of the divisors of n, let $\sigma^*(n)$ and $\sigma^{**}(n)$ be the respective sums of the unitary and bi-unitary divisors of n. Then n is said to be *unitary (bi-unitary) perfect* in case $\sigma^*(n) = 2n$ ($\sigma^{**}(n) = 2n$). The first five unitary perfect numbers are

$$6,\ 60,\ 90,\ 87360,\ 146361946186458562560000.$$

It is not known if there are infinitely many of these numbers.

There are exactly three bi-unitary even perfect numbers and they are 6, 60, and 90.

If N is a unitary perfect number and $N = 2^{\alpha}m$, where m is odd, then the largest prime power unitary divisor of N is called the largest odd *component* of N. Wall [1] showed that, except for the first 5 unitary perfect numbers, the largest odd component must exceed 2^{15}. Later [2, 4] he showed that m must have at least 9 distinct prime factors.

Hansen and Swanson gave a short exposition of some of the basic properties of unitary divisors and proved the two theorems that follow.

Theorem 1. *If a is relatively prime to, and a unitary divisor of, n, then there is a positive integer k such that $k \leq n$ and $a^k \equiv a \pmod n$.*

Theorem 2. *If n is the product of distinct primes, then for all a, $0 \leq a \leq n$, $a^{\phi(n)+1} \equiv a \pmod n$.*

Hagis discussed what is called a *unitary hyperperfect number.* (These are defined in the obvious fashion after the "ordinary" hyperperfect numbers—see Integer 28, Special Perfect Number.)

The above ideas are extended in a very interesting paper by Cohen. In that paper a number d is called a *tri-unitary* divisor of n if the greatest common bi-unitary divisor of n and $\frac{n}{d}$ is 1. In a similar way, one has *quaternary* divisors, etc. Formally, one makes the definition: A divisor d of n is called a *1-ary divisor of* n if the greatest common divisor of d and $\frac{n}{d}$ is 1; and d is called a *k-ary divisor of* n if the greatest common $(k-1)$-ary divisor of n and $\frac{n}{d}$ is 1.

Cohen's first theorem tells us that, for $y \geq 1$ and p a prime, if p^x is a $(y-1)$-ary divisor of p^y, then it is a k-ary divisor of p^y for all $k \geq y-1$.

This result leads to the definition: p^x is an *infinitary divisor* of p^y $(y > 0)$ if p^x is a $(y-1)$-ary divisor of p^y.

We mention only a very few of the results in Cohen's paper.

1. If the binary expansions of x and $y - x$ are, respectively, $\sum x_j 2^j$ and $\sum z_j 2^j$, then p^x is an infinitary divisor of p^y if and only if $\sum x_j z_j = 0$.

2. p^x is an infinitary divisor of p^y if and only if $\binom{y}{x}$ is odd.

3. The only infinitary perfect numbers , i.e., numbers whose infinitary divisors sum to twice themselves, that are not divisible by 8 are the numbers 6, 60, and 90.

Finally, we mention Cohen's interesting "fractal-like" pictures associated with the k-ary divisors of powers of primes. There is a bibliography of 16 items.

See also Integer 6, Perfect Numbers and Integer 28, Special Perfect Number.

REFERENCES

[1] C. R. Wall, On bi-unitary perfect numbers, *Proc. Amer. Math. Soc.* 33 (1972) 39–42.

[2] ——, New unitary perfect numbers have at least nine odd components, *Fib. Quarterly* 26 (1988) 312–317.

[3] R. T. Hansen and L. G. Swanson , Unitary divisors, *Math. Mag.* 52 (1979) 217–222.

[4] C. R. Wall, On the largest odd component of a unitary perfect number, *Fib. Quarterly* 25 (1987) 312–316.

[5] P. Hagis, Jr., Unitary hyperperfect numbers,*Math. Comp.* 36 (1981) 297–98.

[6] G. L. Cohen, On an integer's infinitary divisors, *Math. Comp.* 54 (1990) 395–411.

[7] P. Hagis Jr. and G. L. Cohen, Infinitary harmonic numbers, *Bull. Aust. Math. Soc.* 41 (1990) 151–58.

Integer 96

Prime Subsequence of Primes

Let p_n be the nth prime number and put $q_n = p_{p_n}$. Then every integer greater than 96 is representable as the sum of distinct numbers in the sequence q_n and the 96 may not be replaced by a smaller number.

We append a little data.

n	1	2	3	4	5	6	7	8	9
p_n	2	3	5	7	11	13	17	19	23
q_n	3	5	11	17	31	41	59	67	83

Porubský generalized the above result. Let P_1 be the sequence $p_1, p_2, p_3, \ldots$ of primes, P_2 the sequence $p_{p_1}, p_{p_2}, p_{p_3}, \ldots$ and, in general, P_k be the sequence p_j, where j is in P_{k-1}. Then, for each $k \geq 1$ every sufficiently large integer is a sum of distinct elements taken from P_k.

See also Integer 33, Complete Sequences.

REFERENCES

[1] R. E. Dressler and S. T. Parker , Primes with a prime subscript, *J. Assoc. Comp. Mach.* 22 (1975) 380–81.

[2] Š. Porubský, Sums of distinct terms from a fixed sequence, *Nord. Mat. Tidskr.* 25/26 (1978) 185–87.

Integer 100

1 Fibonacci Mod 100

The nth Fibonacci number u_n ($u_1 = u_2 = 1$) is congruent to n modulo 100 if and only if n is congruent, modulo 60, to one of 1, 5, 25, 29, 41, 49 or is congruent to zero modulo 300. I.e.,

$$u_n \equiv n \pmod{100}$$

if and only if

$$n \equiv 1, 5, 25, 29, 41, 49 \pmod{60} \qquad \text{or} \qquad n \equiv 0 \pmod{300}.$$

A corollary to the stated assertion is: if p is a prime other than 2 or 3, then

$$u_{p^2} \equiv p^2 \pmod{100}.$$

REFERENCES

M. R. Turner, Certain congruence properties (modulo 100) of Fibonacci numbers, *Fib. Quarterly* 12 (1974) 87–91.

2 100 Digits of pi

H. Schubert, in his little book *Mathematical Essays and Recreations,* had the following to say about the accuracy of 100 decimal digits of π.

> An idea can hardly be obtained of the degree of exactness produced by 100 decimal places. But the following example may possibly give us some conception of it. Conceive a sphere constructed with the earth as centre, and imagine its surface to pass through Sirius, which is $134\frac{1}{2}$ millions of millions of kilometres distant from the earth. Then imagine this enormous sphere to be so packed with microbes that in every cubic millimetre millions and millions of these diminutive animalcula are present. Now conceive these microbes to be all unpacked and so distributed singly along a straight line, that every two microbes are as far distant from each other as Sirius from us, that is $134\frac{1}{2}$ million kilometres. Conceive the long line thus fixed by all the microbes, as the diameter of a circle, and imagine the circumference of it to be calculated by multiplying its diameter by π to 100 decimal places. Then, in the case of a circle of this enormous magnitude even, the circumference so calculated would not vary from the real circumference by a millionth part of a millimetre.

Schubert, **not** conceiving of other uses for the decimal digits of π, remarked, "This example will suffice to show that the calculation of π to 100 or 500 decimal places is wholly useless."

Aside from various statistical investigations of the digits of π—questions such as: "Do the 10 digits seem to appear equally frequently." — Borwein, Borwein, and Bailey note that:

> the extended precision calculation of π has substantial application as a test of the "global integrity" of a supercomputer. ... A large-scale calculation of π is entirely unforgiving; it soaks into all parts of the machine and a single bit awry leaves detectible consequences.

Such "excuses" for carrying out calculations which one might very well wish to carry out in any event are not uncommon.

In a digression, we note that Young and Buell [2] said,

As part of a long term test of the hardware reliability of the Cray-2 supercomputer at the supercomputer research center, the authors proved that $F_{20} = 2^{2^{20}} + 1$, which had been the smallest Fermat number of unknown character, is composite.

The authors stated

Some legitimate skepticism must be attached to the conclusion F_{20} is composite, then, since an error either in the hardware or in the computation is almost certainly going to produce an erroneous residue which would nonetheless lead to the expected conclusion that F_{20} is composite. By producing the same residue R_{20} on two different machines, we feel the possibility of hardware errors has been eliminated.

In his review of the paper, Schram observed

The computation to test for the character of F_{20} consists of approximately one million squarings modulo the one million bit number F_{20}. The ultimate output in each case is the residue R_{20} of $3^{(F_{20}-1)/2}$ modulo F_{20}. Recall that F_{20} is prime if and only if $R_{20} = -1$. As the authors note "this may well be the largest computation ever performed whose result is a single bit answer." Because the character of F_{20} is composite, then, the value of this single bit evidently codes up the outcome where $R_{20} \neq -1$. Since there are only two mutually exclusive and exhaustive outcomes consisting of either $R_{20} = -1$ or else $R_{20} \neq -1$, then the joint probability of obtaining $R_{20} \neq -1$ as the indicated output from two machines by chance alone (due, say, to some random malfunction of the circuits) is 0.25; therefore, the authors' conclusion "that the possibility of hardware failure has been eliminated" seems to the reviewer an unwarranted assumption.

Young and Buell also gave the following run down on what is known about the smallest of the Fermat numbers F_n. The left column is the value of n.

0,1,2,3,4	prime
5,6,7,8	composite and completely factored
9,13	one prime factor known, composite cofactor
15,16,17,18,21	one prime factor known, cofactor unknown
10,11	two prime factors known, composite cofactor
19	two prime factors known, cofactor unknown
12	five prime factors known, composite cofactor
14,20	composite, no factor known
22	character unknown

The *New York Times* for June 24, 1990 reported that Arjen Lenstra had completely factored F_9 and that the factors are: 2,424,833; 7,455,602,825,647,884,208,337,395,736,200,454,918,783,366,342,657; and 741,640,062,627,530,801,524,787,141,901,937,474,059,940,781,097,519, 023,905,821,316,144,415,759,504,705,008,092,818,711,693,940,737.

Returning to the theme of computations of the digits of π, it is interesting to note the changes that have occurred over the years. One of the very earliest methods proceeds as follows. Let a_n, b_n be the circumference of the circumscribing, inscribing regular polygons of $6 \cdot 2^n$ sides in a circle of radius $\frac{1}{2}$. Then

$$a_{n+1} = \frac{2a_n b_n}{a_n + b_n}, \qquad b_{n+1} = \sqrt{a_{n+1} b_n}.$$

Starting with $a_0 = 2\sqrt{3}$, $b = 3$ one can, step by step, get better and better approximations for π. By using a method like this Ludolph (1540–1610) calculated π correctly to 34 decimal digits.

From the book by Borwein and Borwein one can construct the following table.

Year	Known digits of π	Time to calculate	Digits per hour
1949	2037	70 hours	29.1
1961	100000	9 hours	11111.1
1973	1000000	24 hours	41666.7
1983	16000000	30 hours	53333.3
1986	29360000	28 hours	1048571.4
1986	33554432	1.6 hours	20971520.0

Borwein and Borwein observed, "In terms of utility, even farfetched applications such as measuring the circumference of the universe require no more digits than Ludolph van Ceulen had available—but then utility has had little to do with this particular story."

For other comments relating to the computation of π see Integer 24, Some Formulae Containing 24.

REFERENCES

[1] H. Schubert, *Mathematical Essays and Recreations,* Open Court, Chicago, 1910 p.140.

[2] D. A. Young and J. Buell, The twentieth Fermat number is composite, *Math. Comp.* 50 (1988) 261–63.

[3] J. M. Schram, *Math. Rev.* 89b:11012

[4] J. M. Borwein, P. B. Borwein, and D. H. Bailey, Ramanujan, modular equations, and approximations to π or How to compute one billion digits of π, *Amer. Math. Monthly* 96 (1989) 201–219.

[5] J. M. Borwein and P. B. Borwein, *Pi and the AGM,* Wiley, New York, 1987.

3 Recursive Sets

For any axiomatizable theory T and any proposition P, if P has a proof in T, then P has another proof consisting of only 243 additions and multiplications of integers. The number 243 was later (1980) reduced to 100. This writer does not know if 100 is best possible.

See also Integer 87, Prime Proof.

REFERENCES

[1] J. P. Jones, Three universal representations of recursively enumerable sets, *J. Symbolic Logic* 43 (1978) 335–351.

[2] ——, Undecidable Diophantine equations, *Bull. Amer. Math. Soc.* 3 (1980) 859–862.

Integer 105

1 Cyclotomic Polynomials

Let $Q_n(x)$ be the monic irreducible integral polynomial with zeros the primitive nth roots of unity. These are called the *cyclotomic polynomials.* If one examines the first few of them, one finds all of the coefficients are either ± 1 or 0. This continues to be the case up to $n = 105$. For $n = 105$ there is a coefficient of 2. (This was first observed by Migotti in 1883.)

In the paper by Vaughan, cited below, there is a table of those cyclotomic polynomials corresponding to odd composite square-free numbers not exceeding 100.

The coefficients do not exceed 2, in absolute value, until $n = 385$.

In 1931 I. Schur proved that, in fact, the set of coefficients of the cyclotomic polynomials of all degrees is not a bounded set. This proof was not published by Schur but is given in the paper by Emma Lehmer cited below.

If $n = 2^\alpha p^\beta q^\gamma$, where all exponents are nonnegative, it is true that all coefficients of Q_n are either ± 1 or 0.

If one lets the nth cyclotomic polynomial be

$$x^n + c_1 x^{n-1} + c_2 x^{n-2} + \cdots + c_{n-1} x + c_n,$$

then in the case of $n = 105$ the coefficient $c_7 = -2$. Endo shows that if, for some n, $|c_k| > 1$, then $k \geq 7$.

There is a very large literature on this subject but we cite only the following four papers.

REFERENCES

[1] G. Kostandi, Une Propriété des équations irréductibles de la division du cercle, *Bull. École Polytech. Bucarest* 14 (1943) 10–18.

[2] R. C. Vaughan, Adventures in arithmetik, or: How to make good use of a Fourier transform, *Math. Intell.* 9 (2) (1987) 53–60.

[3] E. Lehmer, On the magnitude of the coefficients of the cyclotomic polynomials, *Bull. Amer. Math. Soc.* 42 (1936) 389–390.

[4] M. Endo, On the coefficients of the cyclotomic polynomials, *Comment. Math. Univ. St. Paul* 23 (1974/75) 121–26.

2 A Question of Erdős

If $m = 105$, then all of the positive integers of the form $m - 2^k$, with $k \geq 1$, are prime numbers.

Erdős asked if there is an integer larger than 105 for which this is the case?

Uchiyama and Yorinaga defined $P(a, b)$ to be the property possessed by m if for every $k \geq 1$ the quantity $a^k m - b^k$ is prime. The question asked above is just: is 105 the largest integer with property $P(1, 2)$? They gave numerical data concerning the question of whether or not the set of integers having property $P(a, b)$ is always a finite set. For example, they showed that up to $512 \cdot 10^4$ only 4, 6, 12, and 24 have property $P(3, 5)$. They also showed that no integer greater than 105 and smaller than 2^{77} has property $P(1, 2)$.

REFERENCES

[1] P. Erdős, Résultes et problèmes et théorie des nombres, Séminaire Delange-Pisot-Poitou 14e (1972–73) Fasc. 2, Exp. No. 24.

[2] S. Uchiyama and M. Yorinaga, Notes on a conjecture of P. Erdős I (II), *Math. J. Okayama Univ.* 19 (1976/77) 129–140 (20 (1978/79) 41–49).

3 A Decimal Digit Problem

For a given positive integer n, consider the existence of an integer m with the property that m^n has the sum of its decimal digits equal to m.

For each value of n from 1 to 8, the following show the existence of such an m.

$$2^1 = 2, \quad 9^2 = 81, \quad 8^3 = 512, \quad 7^4 = 2401, \quad 28^5 = 17210368,$$
$$18^6 = 34012224, \quad 18^7 = 61222032, \quad 46^8 = 205962976$$

Megill, via Guy, tells us that for every $n \leq 104$ such an m exists but no such m exists for 105.

Thus, 105 is the smallest integer n for which the sum of the base 10 digits in the representation of m^n is different from m for all positive integers m.

REFERENCES

R. K. Guy, The second strong law of small numbers, *Math. Mag.* 63 (1990) 3–20 (Example 61).

4 An Assertion of Cseh

105 is the largest positive integer n for which all smaller odd positive integers relatively prime to n are prime.

REFERENCES

L. Cseh, Generalized integers and Bonse's theorem, *Studia Univ. Babeş—Bolyai Math.* 34 (1989) 3–6.

Integer 113

Functions of Prime Numbers

Let $\psi(x)$ be the natural logarithm of the least common multiple of the integers not exceeding x and let $\theta(x)$ be the natural logarithm of the product of the primes not exceeding x. These functions are well known in the analytic theory of numbers.

In the paper cited below, Rosser and Schoenfeld show, among many other things, that the maximum value of $\frac{\psi(x)}{x}$ is taken for $x = 113$. Further, they show that $\dfrac{\psi(x) - \theta(x)}{\sqrt{x}}$ takes its maximum value at $x = 361$.

REFERENCES

J. B. Rosser and L. Schoenfeld , Approximate formulas for some functions of prime numbers, *Ill. J. Math.* 6 (1962) 64–94.

Integer 117

A Diophantine Equation

In the chapter "Diophantine Equations: p-adic Methods" in *Studies in Number Theory* (edited by W. J. LeVeque), D. J. Lewis states on page 26: "The equation $x^3 - 117y^3 = 5$ is known to have at most 18 integral solutions but the exact number is unknown."

Finkelstein and London (1971) made use of the field $Q(117^{1/3})$, where the cube root is real, to show that, in fact, the equation has no solutions in integers.

Halter-Koch (1973) and Udrescu (1973) independently observed that by considering the equation modulo 9 we get $x^3 \equiv 5 \pmod 9$ and this congruence clearly has no solutions. Consequently we immediately see that the equation has no solutions.

REFERENCES

[1] *Studies in Number Theory* (ed. W. J. LeVeque), Prentice-Hall, New York, 1969, p. 26.

[2] R. Finkelstein and H. London , On D. J. Lewis's equation $x^3 + 117y^3 = 5$, *Can. Math. Bull.* 14 (1971) 111.

[3] F. Halter-Koch, Letter to the editor, *Can. Math. Bull.* 16 (1973) 299.

[4] V. Şt. Udrescu, On D. J. Lewis's equation $x^3 + 117y^3 = 5$, *Rev. Roumaine Math. Pures Appl.* 18 (1973) 473.

Integer 128

Sum of Unequal Squares

128 is the largest integer which is not the sum of unequal squares.

See also Integer 33, Complete Sequences.

REFERENCES

R. Sprague, Über Zerlegungen in ungleiche Quadratzahlen, *Math. Zeit.* 51 (1947–49) 289–290.

Integer 143

Clockhands

In the *New York Review of Books* (August 16, 1990) David Remnick reported the following.

> When the local KGB tried to break his hunger strike in 1985 by dragging him off to the hospital for forced feedings, Sakharov returned, as always, to the consolation of science, easing the pain and boredom with games of pure abstract thought: 'I spent long hours gazing at the clock hanging on the wall of my room. At night, the dim hospital illumination made it at times difficult for me to distinguish the hour hand from the minute hand, and I thought up this brainteaser: An absent minded watch-maker accidentally fastens two hands of equal lengths on a clock with the usual twelve-hour dial. Because of this, there are moments when the time can be read in either of two ways. Find all the ambiguous moments.'

REFERENCES

T. Szirtes, On the problem of interchangeable clock hands, *J. Rec. Math.* 8 (1975/6) 159–168.

Integer 144

Fibonacci Square

144 is the largest square in the Fibonacci sequence.

Changing the initial conditions to $u_1 = 1$, $u_2 = 4$ yields only the squares u_{-8} (= 81), u_1 (= 1), u_2 (= 4), u_4 (= 9). With $u_1 = 1$, $u_2 = 6$ we find as the only squares u_{-2} (= 9), u_1 (= 1), u_{10} (= 225).

In the last two cases, there are no terms of the form $2S^2$.

The *Pell* sequence is defined by $P_0 = 0$, $P_1 = 1$, $P_{n+2} = 2P_{n+1} + P_n$ and its companion sequence by $Q_0 = Q_1 = 1$, $Q_{n+2} = 2Q_{n+1} + Q_n$.

The only squares in the Pell sequence are 1 and 169 and the only square in its companion sequence is 1. The only cube in either sequence is 1.

Wolfskill [6] discusses, in more general terms, squares in second order linear recurrences.

See also Integer 7, Fibonacci Numbers of the Form $k^2 + 1$ and Integer 8, Fibonacci Cubes.

REFERENCES

[1] J. H. E. Cohn, On square Fibonacci numbers, *J. London Math. Soc.* 39 (1964) 537–540.

[2] ——, Lucas and Fibonacci numbers and some diophantine equations, *Proc. Glasgow Math. Assoc.* 7 (1965) 24–8.

[3] A. Eswarathasan, On square pseudo-Fibonacci (Lucas) numbers, *Fib. Quarterly* 16 (1978) 310–314 (*Can. Math. Bull.* 21 (1978) 297–303).

[4] ——, On pseudo-Fibonacci(Lucas) numbers of the form $2S^2$, where S is an integer, *Fib. Quarterly* 17 (1979) 142–437 (*Can. Math. Bull.* 22 (1979) 29–34).

[5] H. V. Krishna, Properties of the Pell sequence, *Math. Education* 5 (1971) 18–20.

[6] J. Wolfskill, Bounding squares in second order recurrence sequences, *Acta Arith.* 54 (1989) 127–145.

Integer 163

1 Approximate Integers

The integer 163 appears in a number of remarkable irrational approximations of integers. In speaking about the area of mathematics in which these sorts of expressions arise, Stark [5] said: "One of the most interesting things about this subject is the wide number of seemingly unrelated items that can be connected by the theory of quadratic fields. Let me give three numerical examples all related to the fact that $h(-163) = 1$." $h(-163)$ is the class number of the quadratic field $Q(\sqrt{-163})$ and his second example is the second one below.

$$\left\{\frac{1}{\pi}\ln(640320^3 + 744)\right\}^2 =$$

$$163.00000000000000000000000000000000232\ldots.$$

$$e^{\pi\sqrt{163}} = 262537412640768743.999999999999250\ldots.$$

$$(e^{\pi\sqrt{163}} - 744)^{1/3} = 640319.9999999999999999999999993903\ldots.$$

A polynomial having a discriminant related to 163 is $x^3 - 8x - 10$. This polynomial has one real root; call it β. Among the first 200 partial quotients in the simple continued fraction expansion of β one exceeds

$16 \cdot 10^6$, another exceeds $1.5 \cdot 10^6$, and several exceed 20000. (In fact, the first few terms of the continued fraction expansion of β are given by

$$\beta = [3, 3, 7, 4, 2, 30, 1, 8, 3, 1, 1, 1, 9, 2, 2, 1, 3, 22986, \ldots].)$$

The discriminant of the cubic is $-4 \cdot (-8)^3 - 27(-10)^2 = -652 = -4 \cdot 163$. Further,

$$\beta + 2 = e^{(1/24)\pi\sqrt{163}} \prod_{n=1}^{\infty} \left\{1 + e^{-(2n+1)\pi\sqrt{163}}\right\}.$$

Finally, $e^{(1/24)\pi\sqrt{163}}$ approximates $\beta + 2$ to 17 decimal places.

See also Integer 24, Some Formulae Containing 24.

REFERENCES

[1] R. F. Churchhouse and S. T. E. Muir, Continued fractions, algebraic numbers and modular invariants, *J. Inst. Math. Appl.* 5 (1969) 318–328.

[2] I. J. Good, What is the most amazing approximate integer in the universe?, *Pi Mu Epsilon J.* 5 (1972) 314–15.

[3] H. M. Stark, A complete determination of the complex quadratic fields of class number one, *Mich. Math. J.* 14 (1967) 1–27.

[4] ——, "An explanation of some exotic continued fractions found by Brillhart" in *Comp. in Number Theory,* A. O. L. Atkin and B. J. Birch (eds.), Proc. Sci. Res. Council Atlas Symp. No. 2, Oxford Univ. Press, Oxford, 1969, pp. 21–35.

[5] ——, "Recent advances in determining all complex quadratic fields of a given class number" in *1969 Number Theory Institute,* D. J. Lewis (ed.), Proc. Symp. Pure Math. vol. xx, Amer. Math. Soc., Providence, 1971, pp. 401–414.

[6] ——, *An Introduction to Number Theory,* Markham, Chicago, 1970, p. 179.

2 Unique Factorization Domains

For $d < 0$ and squarefree, the quadratic field $Q(\sqrt{d})$ has the unique factorization property if and only if d is one of the nine integers

$$-1,\ -2,\ -3,\ -7,\ -11,\ -19,\ -43,\ -67,\ -163.$$

A recent proof of this result has been given by Cherubini and Walliser.

REFERENCES

[1] H. M. Stark, A complete determination of the complex quadratic fields of class number one, *Mich. Math. J.* 14 (1967) 1–27.

[2] J. M. Cherubini and R. V. Walliser, On the computation of all quadratic imaginary fields of class number one, *Math. Comp.* 49 (1987) 295–99.

3 Quadratic Prime Giving Polynomials

The only polynomials (in one indeterminant) that are known to represent infinitely many primes are linear. However, Sierpiński showed in 1964 that if N_n is the number of primes represented by $x^2 + n$, then the sequence $N_1, N_2, \cdots$ is not bounded. This was extended in 1990 to $x^k + n$, $k \geq 2$, by Garrison. It is known that no nonconstant polynomial may represent only primes for large values of its argument. Buck showed that the only rational functions yielding primes for all integer arguments are constants. Generalizing this, Sato and Straus replaced "rational" with "algebraic."

However, there has been some interest in quadratic polynomials from the standpoint of their having many prime values among the smaller values of their arguments.

The quadratic polynomial $x^2 + x + 41$ is prime for all integer values of x in the range $-40 \leq x \leq 39$. This polynomial was discovered by Euler in 1772. The discriminant of this polynomial is -163.

More generally, the following theorem holds. According to Ayoub and Chowla it should be attributed to Frobenius and Rabinovitch (the latter of them having discussed it at the 1912 International Congress of Mathematicians in Cambridge). Unaware of this, Szekeres published a proof in 1974. Ayoub and Chowla give “a very simple proof of the *only if* part using only the most elementary properties of quadratic fields.” The *if* part was proved by Lehmer in 1936. Cohn also gives a proof of both halves of the theorem on p. 156 of his book *A Second Course in Number Theory.*

Theorem. *p is a prime number for which the field $Q(\sqrt{-p})$ has class number 1 if and only if the polynomial*

$$x^2 + x + \frac{p+1}{4}$$

yields only prime values for $0 \le x < \frac{p-3}{4}$.

The Euler polynomial is the special case $p = 163$.

Ribenboim [11] gives a complete proof of this theorem, proceeding almost from first principles. At the beginning of his paper, he relates the following:

> This is the text of a lecture at the University of Rome, on May 8, 1986. The original notes disappeared when my luggage was stolen in Toronto (!); however, I had given a copy to my friend Paolo Maroscia, who did not have his luggage stolen in Rome (!) and was very kind to let me consult his copy. It is good to have friends.

Mollin and Williams [6] have proved the following similar theorem.

Theorem. *If the class number of the field $Q(\sqrt{d})$ is 1 and $f_d(x)$ is given by*

$$f_d(x) = \begin{cases} -x^2 + x + \frac{d-1}{4} & \text{if } d \equiv 1 \pmod 4; \\ d - x^2 & \text{otherwise,} \end{cases}$$

then $f_d(x)$ is prime for all integers x such that $1 < x < \alpha$, where

$$\alpha = \begin{cases} \frac{1}{2}\sqrt{d-1} & \text{when } d \equiv 1 \pmod 4; \\ \sqrt{d} & \text{otherwise.} \end{cases}$$

They give some data on particularly good polynomials. For example, the polynomials

$$3x^2 + 3x - 89, \quad -2x^2 + 2x + 113, \quad 2x^2 - 199$$

yield either ± 1 or a prime for, respectively, 3515, 3585, 4373 values of $x \leq 10000$. (The Euler polynomial yields 4149 primes in this range.)

One consequence of this and the result mentioned in the preceding subsection is that for no integer A larger than 41 can the polynomial $x^2 + x + A$ be prime for all x satisfying $0 \leq x \leq A - 2$. In fact, the only prime A for which the given quadratic is prime for all x such that $0 \leq x \leq A - 2$ are $A = 2, 3, 5, 11, 17$, and 41.

Fendel connects such polynomials to unique factorization domains and proves the following theorems. In these statements, the symbol D_n stands for the ring of algebraic integers in the field $Q(\sqrt{-n})$ and $n = 4C - 1$.

If D_n is a unique factorization domain, then $x^2 + x + C$ is prime for all x satisfying $0 \leq x \leq C - 2$.

If $x^2 + x + C$ is prime for all x satisfying $0 \leq x \leq [\sqrt{\frac{n}{12}}]$, then D_n is a principal ideal domain.

If $n \equiv 3 \pmod 4$ and D_n is a unique factorization domain, then the quadratic is prime for $0 \leq x \leq C - 2$.

If n is congruent to 1 or 2 modulo 4 and is larger than 2, then D_n is not a unique factorization domain.

Beeger gave the polynomial $x^2 + x + 72491$ which yields primes for 4923 of the first 11000 values of x. In contrast, the polynomial $x^2 + x + 41$ yields 4506 primes in this range.

In a review of a paper by Mařik (*Math. Rev.* 18, 16d), Lehmer says that the author, in quoting a previous theorem, asserts "if p is a prime for which $x^2 + x + p$ is also prime for all x for which $x^2 + x \leq \frac{p-1}{3}$, then every number $< p^2$ properly represented by the form $x^2 + xy + py^2$ is a prime." This is said to be "closely related" to a 1912 result of Frobenius. The hypothesis of the assertion is stated to be true for $p = 2, 3, 5, 11, 17$, and 41.

Karst carried out a numerical investigation of polynomials of the form $Ax^2 + Ax - C$ for $A < 10$ and $C < 2 \cdot 10^5$. Some of his data is listed in the table below. (n_{icp} is the number of initial consecutive primes.)

Form and n_{icp}		Number of prime $f(x)$ for $x <$							
Form	n_{icp}	100	200	300	400	500	600	700	800
$x^2 + x + 41$	40	86	156	211	270	326	383	431	479
$2x^2 - 199$	28	88	150	216	273	332	382	445	493
$9x^2 + 9x + 43$	13	69	124	177	222	263	303	347	395
$8x^2 + 8x - 197$	31	77	138	193	247	295	345	402	455

Fung and Williams give further results of a similar kind in a recent issue of *Mathematics of Computation.*

Higgins reports that the two polynomials

1. $9x^2 - 231x + 1523$
2. $9x^2 - 471x + 6203$

yield distinct primes, in reverse order from each other, for the values $0 \leq x \leq 39$. The discriminant of each of these polynomials is $-9 \cdot 163$.

Letting p_j be the jth prime number it is clear, by passing a polynomial of degree $m - 2$ through the $m - 1$ points

$$(1, 2), (2, 3), \ldots, (m - 1, p_{m-1}),$$

that there exist polynomials of degree $m - 2$ yielding only primes for the initial $m - 1$ positive integer values of their arguments.

Chang and Lih proved this theorem, in the case where m is prime, using Dirichlet's theorem on primes in arithmetic progressions. They were not able to make any general statement concerning the minimal degree for such a polynomial by their method other than that it cannot exceed $m - 1$.

However, their method gives rise to the following (mostly quadratic) polynomials $f_p(x)$, where $f_p(x)$ is prime for $0 \leq x < p$.

p	$f_p(x)$
2	$x + 2$
3	$2x + 3$
5	$2x^2 + 5$
7	$2x^2 - 2x + 7$
11	$x^2 - x + 11$
13	$6x^2 + 13$
17	$x^2 - x + 17$
19	$2x^2 - 2x + 19$
23	$3x^2 - 3x + 23$
29	$2x^2 + 29$

For $p = 31$ they were not able to use their method to get a quadratic polynomial of the appropriate kind.

Ribenboim mentions the problem of finding arithmetic progressions of prime numbers; in particular, progressions with typical term $ax + b$, where one wishes a prime for each of the values $x = 0, 1, \ldots, b-1$. For $b = 5, 7, 11, 13$, he gives the following.

$$6x + 5$$
$$150x + 7$$
$$1536160080x + 11$$
$$9918821194590x + 13$$

(See Integer 30, A Spurious Property, for other information on primes in arithmetic progression.)

In a slightly different direction, Iwaniec (1978) proved the following theorem.

Theorem. *If $G(n) = an^2 + bn + c$ is an irreducible integral polynomial with $a > 0$ and c odd, then there are infinitely many n such that $G(n)$ has at most two prime factors. As a special case we see that $n^2 + 1$ is a product of at most two prime factors infinitely often.*

In 1916 Brun had shown that the quotient of the number of $n^2 + 1$ primes not exceeding x by the total number of $n^2 + 1$ numbers not exceeding x has a limit of 0 as x tends to infinity.

In 1981 Betty Garrison showed that there are arbitrarily long strings of consecutive n for which $n^2 + 1$ is composite.

A mildly interesting unrelated fact about quadratic polynomials is the following due to Allison. An integral polynomial which is not a square may take square values for 8 consecutive integer values of its argument. In fact an infinite family of such polynomials is given.

REFERENCES

[1] W. Sierpiński, Les binôme $x^2 + n$ et nombres premieres, *Bull. Soc. Royale Sci. Liège* 33 (1964) 259–260.

[2] B. Garrison, Polynomials with large numbers of prime values, *Amer. Math. Monthly* 97 (1990) 316–17.

[3] R. C. Buck, Prime-representing functions, *Amer. Math. Monthly* 53 (1946) 265.

[4] D. Sato and E. G. Straus, p-adic proof of non-existence of proper prime representing algebraic functions and related problems, *J. London Math. Soc.* 2 (1970) 45–48.

[5] R. G. Ayoub and S. Chowla, On Euler's polynomial, *J. Number Theory* 13 (1981) 443–45.

[6] R. A. Mollin and H. C. Williams, Prime producing polynomials and real quadratic fields of class number one in *Number Theory,* J.-M. Dekoninck and C. Levesque (eds.), de Gruyter, New York, 1989, pp. 654–663.

[7] ——, Class number one for real quadratic fields, continued fractions and reduced ideals in *Number Theory and Applications,* R. A. Mollin (ed.), Kluwer, Dordrecht, 1988, pp. 481–496.

[8] D. Fendel, Prime-producing polynomials and principal ideal domains, *Math. Mag.* 58 (1985) 204–210.

[9] N. G. W. H. Beeger, Report on some calculations of prime numbers, *Nieuw Arch. Wisk.* 20 (1939) 48–50.

[10] G. Rabinovitch, Eindeutigkeit der Zerlegung in Primzahlfacktoren in quadratischen Zahlenkörpern in *Fifth International Congress of Mathematicians Proceedings,* Cambridge, vol.1, 1912, pp. 418–421; (also *J. für Math.* 142 (1913) 153–164).

[11] P. Ribenboim, Euler's famous prime generating polynomial and the class number of imaginary quadratic fields, *L'Ens. Math.* 34 (1988)

23–42.

[12] G. Szekeres, On the number of divisors of $x^2 + x + A$, *J. Number Theory* 6 (1974) 434–442.

[13] E. Karst, New quadratic forms with high density of primes, *Elem. Math.* 28 (1973) 116–118.

[14] G. W. Fung and H. C. Williams, Quadratic polynomials which have a high density of prime values, *Math. Comp.* 55 (1990) 345–353.

[15] O. Higgins, Another long string of primes, *J. Rec. Math.* 14 (1981/2) 185.

[16] G. J. Chang and K. W. Lih, Polynomial representation of primes, *Tamkang J. Math.* 8 (1977) 197–98.

[17] H. Iwaniec, Almost-primes represented by quadratic polynomials, *Invent. Math.* 47 (1978) 171–188.

[18] B. Garrison, Consecutive integers for which $n^2 + 1$ is composite, *Pac. J. Math.* 97 (1981) 93–96.

[19] W. Allison, On square values of polynomials, *Math. Colloq. U. Cape Town* 9 (1974) 135–141.

[20] ——, On certain simultaneous Diophantine equations, *Math. Colloq. U. Cape Town* 11 (1977) 117–133.

Integer 239

Sums of Cubes

Every positive integer may be written as a sum of no more than 9 cubes and every sufficiently large integer may be written as a sum of no more than 7 cubes. It is not known if the 7 may be diminished. However, there are exactly two numbers which require 9 cubes and they are 23 and 239.

The integer 7 may not be written as a sum of fewer than 4 squares: $7 = 2^2 + 3 \cdot 1^2$. As noted above 23 requires 9 cubes: $23 = 2 \cdot 2^3 + 7 \cdot 1^3$. Similarly 79 requires 19 4th powers and 223 requires 37 5th powers: $79 = 4 \cdot 2^4 + 15 \cdot 1^4$, $223 = 6 \cdot 2^5 + 31 \cdot 1^5$.

Numerical evidence seems to indicate 454 is the largest integer requiring 8 cubes and 8042 the largest requiring 7 cubes.

There is an integer $G(k)$, $k \geq 2$, such that all sufficiently large integers may be expressed as a sum of $G(k)$ or fewer kth powers. The existence of this number guarantees the existence of a number $g(k)$ such that **every** integer is a sum of $g(k)$ kth powers and some integer requires the use of $g(k)$ such powers.

Maillet and Hurwitz show that $G(k) \geq k + 1$ and a quite elementary argument shows

$$g(k) \geq 2^k + \left[\left(\frac{3}{2}\right)^k\right] - 2, \qquad k \geq 1. \tag{1}$$

In fact, by a theorem of Mahler , proved in 1957, the last inequality is an equality except for a finite number of integers.

Prior to Mahler's work it was known that if

$$3^k - 2^k \left[\frac{3}{2}^k\right] < 2^k - \left[\frac{3}{2}^k\right],$$

then the first inequality in equation 1 is an equality whereas if this inequality is reversed that inequality is strict.

As an aside we would like to mention that though it is known that, for almost all (in the Lebesgue measure sense) real $\alpha > 1$, $[\alpha^n]$ is composite for infinitely many n it is generally quite difficult to lay hands on a specific α for which this is true. Forman and Shapiro show this to be the case for α each of $\frac{4}{3}$ and $\frac{3}{2}$.

See also Integer 19, Waring's Problem.

REFERENCES

[1] K. Chandrasehkaran, Exponential sums in the development of number theory, *Proc. Steklov Inst.* 132 (1973) 7–26, 264.

[2] L. E. Dickson, All integers except 23 and 239 are sums of eight cubes, *Bull. Amer. Math. Soc.* 45 (1939) 588–591.

[3] C. Small, Waring's problem, *Math. Mag.* 50 (1977) 12–16.

[4] K. Mahler, On the fractional parts of the powers of a rational number(II), *Mathematika* 4 (1957) 122–24.

[5] W. Forman and H. N. Shapiro, An arithmetic property of certain rational powers, *Comm. Pure and Appl. Math.* 20 (1967) 561–573.

Integer 283

Triplets for Primes

Three consecutive positive integers constitute a *triplet* if for some prime number p each of the three numbers is a cubic residue of p. A prime number is called *exceptional* if it has no triplets. The following three statements are true.

a. The only exceptional primes are:

$$2, 3, 7, 13, 19, 31, 37, 43, 61, 67, 79, 127, 283;$$

b. Every nonexceptional prime has a triplet that does not exceed

$$(23532,\ 23533,\ 23534);$$

c. The triplet in (b) is best possible since there are infinitely many primes having that as smallest triplet.

With respect to quadratic residues, Brauer proved, in 1928, that given a positive integer m every sufficiently large prime has a run of m consecutive quadratic residues. If we denote by $r(2, m, p)$ the least r such that $r, r+1, r+2, \ldots, r+m-1$ are all quadratic residues of p, then by a 1962 result of Lehmer and Lehmer the function is unbounded for $m \geq 3$. In 1973 Hudson showed $r(2, m, p) > c \log p$ for some constant c and infinitely many primes p.

REFERENCES

[1] D. H. Lehmer, E. Lehmer, W. H. Mills, and J. L. Selfridge, Machine proof of a theorem on cubic residues, *Math. Comp.* 16 (1962) 407–415.

[2] R. H. Hudson, A note on Dirichlet characters, *Math. Comp.* 27 (1973) 973–75.

Integer 495

Kaprekar Constants

Let a be an r-digit number in base 10 for which not all digits are equal. Let $m(a), M(a)$ be the least and greatest integers obtainable from a by permuting its digits. If we put $T(a) = M(a) - m(a)$, then, providing there are three digits in a and they are not all the same, iteration of T ultimately leads to the integer 495. Since $T(495) = 495$ the process ceases to produce new numbers at this stage.

The integer 495 is the only three-digit number having this property. An integer a is called a *Kaprekar constant* if iteration of T always leads to a and $T(a) = a$. Thus there is only one three-digit Kaprekar constant.

In fact, Pritchett, Ludington, and Lapenta have shown that there are only two decadic Kaprekar constants altogether and they are 495 and 6174.

That 6174 is such a constant was apparently first observed by Kaprekar [2] in 1949.

Four-digit Kaprekar constants can exist only in bases g where $g = 2^n \cdot 5$, where n is either 0 or odd. In each case there is such a constant and if we put $a = 2^n \cdot 3,\ b = 2^n$ the Kaprekar constant is:

$$a(b-1)(g-b-1)(g-a) \qquad \text{if } 0 < b < a;$$

$$a(a-1)(g-a-1)(g-a) \qquad \text{if } 0 < a = b;$$

$$(a-1)(g-1)(g-1)(g-a) \qquad \text{if } 0 = b < a.$$

When $g = 10$ then $n = 1$ so $a = 6, b = 2$. This puts us in the first case so the constant is 6174.

Recently, Trigg defined a similar operation on four-digit base ten integers for which all digits are not equal. If $M(n)$ is as before and is equal to $abcd$, $a \geq b \geq c \geq d$, define f by $f(n) = badc - cdab$. Repeated application of f leads to 2538, which goes to itself. For example, starting with $n = 7162$ we find $a = 7, b = 6, c = 2, d = 1$ (since $M(n) = 7621$) so

$$f(7162) = 6712 - 2176 = 4536.$$

Continuing

$$f(4536) = 5634 - 4365 = 1269,$$
$$f(1269) = 6912 - 2196 = 4716,$$
$$f(4716) = 6714 - 4176 = 2538.$$

Kaprekar [4] has introduced another sort of number which he has also called a *Kaprekar number.* These are numbers having the property exhibited by the examples:

$$45^2 = 2025,\ 20 + 25 = 45; \qquad 297^2 = 88209,\ 88 + 209 = 297.$$

He asserts that the smallest such number having ten digits is 1111111111.

$$1111111111^2 = 1234567900987654321$$
$$123456790 + 0987654321 = 1111111111.$$

REFERENCES

[1] G. D. Pritchett, A. L. Ludington, and J. F. Lapenta, The determination of all decadic Kaprekar constants, *Fib. Quarterly* 19 (1981) 45–52.

[2] D. R. Kaprekar, Another solitaire game, *Scripta Math.* 15 (1949) 244–45.

[3] H. Hasse and G. D. Pritchett, The determination of all four-digit Kaprekar constants, *J. für Reine und Angew. Math.* 299/300 (1978) 113–124.

[4] D. R. Kaprekar, On Kaprekar numbers, *J. Rec. Math.* 13 (1980/81) 81–2.

[5] K. E. Eldridge and S. Sagong, The determination of Kaprekar convergence and loop convergence of all three digit numbers, *Amer. Math. Monthly* 95 (1988) 105–112.

[6] C. W. Trigg, A new routine leads to a new constant 2538, *J. Rec. Math.* 12 (1979-80) 209–210.

Integer 561

Carmichael Numbers

A *Carmichael* number is a positive composite integer n such that

$$a^{n-1} \equiv 1 \pmod{n}$$

for every integer a prime to n.

The smallest Carmichael number is 561.

The number 62745 is the only Carmichael number of the form $15pq$. There are no such numbers of the forms $21pq$ or $39pq$. (The p, q here are primes.)

A Carmichael number must be a product of at least three distinct odd primes such that each of them when diminished by 1 divides $n - 1$.

There are various characterizations of classes of Carmichael numbers. One (due to Chernick) is as follows: if

$$n = (6m + 1)(12m + 1)(18m + 1)$$

and the right-hand factors are all prime, then n is a Carmichael number. Using a modification of this, Dubner has found some quite large examples.

Schinzel showed that for a given natural number a there are arbitrarily large primes p, q such that pq is a divisor of $a^{pq} - a$. This was extended later, by both him and by Rotkiewicz, to products of more than two primes. In fact, Rotkiewicz showed that if a and b are relatively prime and s is an integer, then there are infinitely many n, each the product of s distinct primes, such that n is a divisor of $a^{n-1} - b^{n-1}$.

Schinzel put m_a for the least composite n for which $n|a^n - a$. He observed that $m_1 = 4$, $m_2 = 341$, $m_{4k} = m_{4k+1} = 4$, $m_{12k+3} = m_{12k+6} = m_{12+7} = m_{12k+10} = 6$ and that each value taken on by any m_a is taken on infinitely often. Because 561 is a Carmichael number the sequence

$$m_1, m_2, m_3, \ldots$$

ceases, after the 561st term, to have any terms that have not previously appeared.

Jaeschke tabulated the known Carmichael numbers up to 10^{12}.

See also Integer 23, Pseudoprimes.

REFERENCES

[1] R. D. Carmichael, On composite numbers P which satisfy the Fermat congruence $a^{P-1} \equiv 1 \pmod{P}$, *Amer. Math. Monthly* 19 (1912) 22–7.

[2] H. J. A. Duparc, On Carmichael numbers, *Simon Stevin* 29 (1952) 21–4.

[3] H. Dubner, A new method for producing large Carmichael numbers, *Math. Comp.* 53 (1989) 411–414.

[4] H. Dubner and H. Nelson, Carmichael numbers which are the product of three Carmichael numbers, *J. Rec. Math.* 22 (1990) 2–6.

[5] A. Schinzel, Sur les nombres composés n qui divisent $a^n - a$, *Rend. Circ. Mat Palermo* 7 (1958) 37–41.

[6] A. Rotkiewicz, Sur les nombres n qui divisent $a^{n-1} - b^{n-1}$, *Rend. Circ. Mat. Palermo* 8 (1959) 115–16.

[7] G. Jaeschke, The Carmichael numbers to 10^{12}, *Math. Comp.* 55 (1990) 383–89.

Integer 563

Wilson Remainders

In elementary number theory it is proved that if p is a prime, then $(p-1)!+1$ is divisible by p. Let q denote the quotient of $(p-1)!+1$ by p. *Wilson's remainder,* denoted W_p, is the least nonnegative remainder under division of q by p. When $W_p = 0$, i.e., when

$$(p-1)! \equiv -1 \pmod{p^2},$$

p is called a *Wilson prime.*

A number of people have been interested in whether or not an infinity of Wilson primes exists. Fröberg has shown that for $3 \leq p < 50000$ the only Wilson primes are 5, 13, 563. The third Wilson prime 563 was first found by Goldberg who gave a table of the Wilson quotients up to 10000. Later, Pearson showed the nonexistence of Wilson primes in the range $30000 < p \leq 200183$.

REFERENCES

[1] C.-E. Fröberg, Investigation of the Wilson remainders in the interval $3 \leq p < 50000$, *Ark. Math.* 4 (1963) 479–499.

[2] K. Goldberg, A table of Wilson quotients and the third Wilson prime, *J. London Math. Soc.* 28 (1953) 252–56.

[3] E. H. Pearson, On the congruences $(p-1)! \equiv -1$ and $2^{p-1} \equiv 1 \pmod{p^2}$, *Math. Comp.* 17 (1963) 194–95.

Integer 645

Sum of Consecutive Integers

The integer 645 is the largest integer such that the sum of the positive integers not exceeding it is a sum of squares of consecutive integers. In fact,

$$1 + 2 + \cdots + 645 = 1^2 + 2^2 + \cdots + 85^2.$$

The only other solutions to

$$1 + 2 + \cdots + n = 1^2 + 2^2 + \cdots + r^2$$

are $(n, r) = (1, 1)$, $(10, 5)$, and $(13, 6)$. The sum of squares of consecutive integers, starting with 1, is itself a square only when the number of integers involved is 24. (See Integer 24, Consecutive Sum of Squares a Square.)

See also Integer 4, A Few Diophantine Equations.

REFERENCES

[1] È. T. Avanesov, The Diophantine equation $3y(y + 1) = x(x + 1)(2x + 1)$, *Volž. Mat. Sb. Vyp.* 8 (1971) 3–6.

[2] R. Finkelstein, H. London, On triangular numbers which are sums of consecutive squares, J. Number Theory 4 (1972) 455–462.

Integer 654

Gaps in Primes

Though it is clear that there exist arbitrarily long strings of consecutive composite integers, e.g.,

$$n! + 2, n! + 3, \ldots, n! + n$$

is a string of $n - 1$ consecutive composite integers, it is difficult to come by strings of any great length without going to excessively large integers.

In 1981 Weintraub exhibited a gap of length 654 following the prime number 11 000 001 446 613 353. The previous "largest" gap was of length 652.

REFERENCES

S. Weintraub, A large prime gap, *Math. Comp.* 36 (1981) 279.

Integer 691

A Divisibility Property

The τ function, see Integer 23, The τ Function, and Integer 63001, Ramanujan's τ Function, satisfies the following congruence.

$$\tau(n) \equiv 0 \pmod{691} \qquad \text{for almost all } n.$$

Even though 691 divides $\tau(n)$ for almost all n the smallest such n is 1381.

The 691 in this congruence may be replaced by $2^5 \cdot 3^3 \cdot 5^2 \cdot 7^2 \cdot 23 \cdot 691$.

But in 1974 Serre showed that for any fixed integer m the following is true.

$$\tau(n) \equiv 0 \pmod{m} \qquad \text{for almost all } n.$$

Does this not vitiate the above "interesting" property of 691? Only a close examination of the history (and romance?) of this particular branch of mathematics will enable one to answer the question for oneself.

Possibly to reinstate 691 we note that

$$\tau(n) \equiv \sum_{d|n} d^{11} \pmod{691}.$$

A related result, due to Pančiškin, is that, for r, s arbitrary integers prime to 691, the number of primes p such that $p \equiv r \pmod{691}$ for which

$$\tau(p) \equiv p^n + 1 + 691s \pmod{691^2}$$

has positive density.

Chowla notes that the sum $\sum_{d|n} d^{11}$ also appears in the following context.

Let $r_{24}(n)$ be the number of representations of n as a sum of 24 squares. Then there are fixed integers A, B, independent of n, such that

$$r_{24}(n) = A \sum_{d|n} d^{11} + B\tau(n).$$

See the articles by Rankin, Murty, and Swinnerton-Dyer in *Ramanujan Revisited.*

As a final observation we note that in 1973 Rankin showed that

$$\frac{1}{x} \sum_{p \leq x} \frac{\tau^2(p) \log p}{p^{11}} \to 1 \qquad \text{as } x \to \infty.$$

In 1984 Grupp generalized this to give

$$\frac{1}{x} \sum_{p \leq x, p \equiv \ell \pmod{q}} \frac{\tau^2(p) \log p}{p^{11}} \to \frac{1}{\phi(q)} \qquad \text{as } x \to \infty,$$

for ℓ and q relatively prime.

See also Integer 23, The τ Function and Integer 63001, Ramanujan's τ Function.

REFERENCES

[1] *Ramanujan Revisited,* G. E. Andrews, et al. (eds.), Academic Press, New York, 1988, pp. 245–311.

[2] A. A. Pančiškin, Ramanujan congruences mod 691^2 do not exist (Russian), *Mat. Zametki* 17 (1975) 255–264 (English trans. *Math. Notes* 17 (1975) 148–153).

[3] S. Chowla, *The Riemann Hypothesis and Hilbert's Tenth Problem,* Gordon and Breach, New York, 1965, p. 59.

[4] F. Grupp, Eine Bemerkungen zur Ramanujan'schen τ-Funktion, *Arch. Math.* 43 (1984) 358–363.

Integers 714 and 715

714 and 715

Let P_k be the product of the first k prime numbers. Then we have

$$P_1 = 1 \cdot 2, \quad P_2 = 2 \cdot 3, \quad P_3 = 5 \cdot 6, \quad P_4 = 14 \cdot 15, \quad P_7 = 714 \cdot 715.$$

That is, each of P_1, P_2, P_3, P_4, P_7 is a product of two consecutive integers. There are no other k for which this is true of P_k for $k \leq 3049$. That is, for P_k, with $k > 7$ to have the stated property it must be the case that P_k exceeds 10^{6021}, a rather large integer.

Put $S(n) = a_1 p_1 + \cdots + a_k p_k$ when $p_1^{a_1} \cdots p_k^{a_k}$ is the canonical prime decomposition of n. (Compare with Integer 7, Iterates.)

The function S satisfies $S(mn) = S(m) + S(n)$.

Also $S(714) = S(715)$, $S(\sigma(714)) = S(\sigma(715))$, $\sigma(714)$ is a cube, $\sigma(714)/\phi(714)$ is a square, $\phi(\sigma(714)) = 2\phi(\sigma(715))$ is a square, the sum of 714 and 715 is 1429 and all of the numbers 1429, 9241, 1249, 9421, 4129, 4219 are primes.

See also Integer 5, Numbers of the Form $n! + k$.

REFERENCES

C. Nelson, D. E. Penney, and C. Pomerance, 714 and 715, *J. Rec. Math.* 7 (1974) 87–89.

Integer 720

Factorials

The highest power of the prime p which divides $p!$ is 1. Among the first 11 integers n the highest power of n which divides $n!$ is 1 for $n = 2, 3, 4, 5, 7$ and is 2 for $n = 6, 8, 9, 10$. For $n = 12$ the highest power is 5.

It seems that the highest power is generally not too large. MacKinnon discusses a reasonably sharp estimate for this highest power and observes that the highest power of 720 dividing 720! is 178. Thus 720 is one of the numbers for which this highest power is "fairly large."

Two other examples are:

$$1728^{287} | 1728!$$

and

$$518400^{64798} | 518400!.$$

Metelka discusses the existence of n such that for given $m > 1$, $k \geq 1$, $m^k | n!$ but $m^{k+1} \not| n!$. For m prime there are either no solutions at all or there are m consecutive integer solutions.

As an example for $m = 7$, $k = 53$ there are m solutions for n, namely, $329, 330, \ldots, 335$.

REFERENCES

[1] N. MacKinnon, How many times does n go into $n!$, *Math. Gaz.* 70 (1986) 203–205.

[2] J. Metelka, On the divisors of the factorial (Czech.), *Mat.-Fyz. Časopis Sloven. Akad. Vied* 15 (1965) 60–72.

Integer 1093

Fermat's Conjecture

Fermat's conjecture says that for $n > 2$ there are no nontrivial solutions for x, y, z in the equation

$$x^n + y^n = z^n.$$

Though stating (in his copy of a book by Diophantus) the truth of this conjecture in 1637 with a remark, quoted by Dickson in vol. 2 of his *History of the Theory of Numbers*,

> I have discovered a truly remarkable proof which this margin is too small to contain

the result has still not been completely proved. The conjecture often goes under the name of Fermat's last theorem (FLT).

It is not difficult to see that the conjecture can be reduced to the conjecture that the equation has no nontrivial solutions when n is a prime number.

In the history of the subject it has become common to say that the *first case* of the theorem *holds* for the prime p if the existence of integers $x, y, z,$

$$x^p + y^p = z^p,$$

implies that p divides the product xyz.

In 1909 Wieferich proved that if the first case of FLT **fails** for the exponent p, then p satisfies the congruence

$$2^{p-1} \equiv 1 \pmod{p^2}.$$

(Such primes are sometimes referred to as *Wieferich squares.*) Consequently, if one could show, for a given prime p, that this congruence was false, then one would know that any solution would necessarily have $p|xyz$. E.g., since $2^{3-1} \not\equiv 1 \pmod{3^2}$ we see that if $x^3 + y^3 = z^3$, then $3|xyz$.

This led to the question of whether such primes p for which $2^{p-1} \equiv 1 \pmod{p^2}$ existed. The first such prime, 1093, was found in 1913 by Meissner. In 1922 Beeger found the second such known prime, 3511. It is now known that no other such primes exist below $6 \cdot 10^9$. (See Lehmer.)

(See Integer 23, Pseudoprimes, for another appearance of 1093 and 3511.)

In 1910 Miramanoff showed:

If the first case fails for p, then $3^{p-1} \equiv 1 \pmod{p^2}$.

Direct checking shows that neither 1093 nor 3511 satisfies this congruence. Thus Wieferich's and Miramanoff's criteria together show that for all primes up to $6 \cdot 10^9$ the first case holds.

The only primes p less than 2^{30} for which $3^{p-1} \equiv 1 \pmod{p^2}$ are 11 and 1006003.

In 1914 Frobenius and Vandiver showed that failure of the first case for p implies that both of the congruences $5^{p-1} \equiv 1 \pmod{p^2}$ and $11^{p-1} \equiv 1 \pmod{p^2}$ hold.

Later Pollaczek, Vandiver, and Morishima showed that failure of the first case for a prime p implies $m^{p-1} \equiv 1 \pmod{p^2}$ for all prime m, $m \leq 31$.

Granville and Monagan replaced the 31 by 89 in 1988. Thus they were able to show that the first case of FLT is true for all prime exponents up to 714 591 416 091 389. Writing $(FLTI)_p$ for the first case of the Fermat theorem for the prime p they say "... Shanks and Williams observed that if one could show that $(FLTI)_p$ is false implies p^2 divides $q^p - q$ for each $q \leq 109$, then $(FLTI)_p$ is true for $p < 4408660978137503$. Although this was our initial objective, we were only able to complete the computation as far as $q = 89$, and so prove the theorem stated in the title."

The above bound was raised to $7.568 \cdot 10^{17}$ by Coppersmith in 1990.

Adelman and Heath-Brown have shown that the first case is true for infinitely many p. (Granville, connecting the first case to the theory of "powerful numbers," proved the same thing based on the unproved conjecture that there do not exist three consecutive "powerful numbers." A *powerful number* is an integer whose canonical prime factorization has no exponents equal to unity. In his 1988 paper, Ribenboim [8] discussed this and various other related conjectures.)

In 1985 Heath-Brown showed that FLT is true for "almost all exponents" in the following sense. If $N(x)$ is the number of n not exceeding x for which the Fermat conjecture is false, then the quotient $N(x)/x$ has the limit 0 as x tends to infinity.

In another sense of "almost all" it can be shown that the following conjecture of Masser and Oesterlé implies Fermat's last theorem for all sufficiently large exponents.

(*abc-conjecture*) Let a, b, c be integers whose gcd is 1 and whose sum is 0. Then for any $\epsilon > 0$ there is a constant $C(\epsilon)$, depending only upon ϵ, such that

$$\max\{|a|, |b|, |c|\} < C(\epsilon) \left(\prod_{p|abc} p \right)^{1+\epsilon}.$$

In fact, given the *abc*-conjecture, if $x^n + y^n = z^n$, then (taking $a = x^n$, $b = y^n$, $c = z^n$) we see that

$$\max\{|x|, |y|, |z|\}^n < C(\epsilon) \left(\prod_{p|xyz} p \right)^{1+\epsilon} < C(\epsilon) \max\{|x|, |y|, |z|\}^{3+3\epsilon}$$

which is clearly not true for large n.

In a review, Hirzebruch said $2^{p-1} \equiv 1 \pmod{p^2}$ holds "almost certainly for infinitely many primes" and $2^{p-1} \equiv 1 \pmod{p^3}$ "holds for no prime $< 3 \cdot 10^9$ and almost certainly for no prime at all."

In connection with this last congruence, Ribenboim had the following to say.

> In the literature there is a paper by Linkovski(1968), in which he claims: If the first case of Fermat's theorem fails for p, then $2^{p-1} \equiv 1 \pmod{p^3}$. This would represent an outstanding strengthening of Wieferich's theorem. However, Linkovski's proof is not correct, since it is based on ... an "assertion published by Grebeniuk in 1956 ..." and "In 1975 Gandhi and Stuff analyzed Grebeniuk's proof and found a mistaken deduction ..."

The interested reader should consult the book by Ribenboim [13, p. 153] for further information about this—including the mention of a correct deduction of FLT based on Linkovski's incorrect result.

In a review of two papers by H. M. Edwards, W. C. Waterhouse said the following.

> There is a well known story that Kummer thought he could prove Fermat's last theorem, but Dirichlet pointed out to him that he was assuming unique factorization of cyclotomic integers. The author traces this story to a memorial lecture sixty years later in which Hensel says only that it comes from "unimpeachable" sources. He also argues that Kummer's familiarity with the reciprocity work of Jacobi and Gauss makes the story unlikely. Lamé, however, did make precisely this mistake (and Liouville at once corrected him).
>
> Looking for the truth behind the story, the author found that there is a gap in Kummer's 1847 proof for regular primes. ... This was caught by Cauchy and Liouville in 1857 when they were preparing to award Kummer the Paris Academy prize; independently he caught it himself and published a correct proof.
>
> But there is also an 1845 letter to Eisenstein mentioning some error that led to actual false results related to unique

> factorization. With the indications this gave of where to look, the author found that the Berlin Academy still has on file a short 1844 paper withdrawn by Kummer before publication. ... The error in the proof is that, having two elements with no common factor, Kummer assumes that every element can be written as a combination of them. A letter from Jacobi to Dirichlet shows that it was Jacobi who recognized the claim was wrong and thereby prompted Kummer to do the computations illustrating nonunique factorization.

For a complete history of the problem see the book by Ribenboim [13] cited below. It is now known that Fermat's conjecture is true for, at least, all primes not exceeding 150000. (See Tanner and Wagstaff.)

It is amusing to note that although the Fermat conjecture remains unsolved there are "easy" problems that might, superficially, seem to be of equal difficulty. One such problem is to solve the equation

$$x^n + y^n = z^{n+1}.$$

One might think of Watson, not seeing how to solve the problem, assuring Sherlock Holmes that he will not say, "Oh, how very simple!" but almost instantly forgetting himself when Holmes points out that

$$(a(a^n + b^n))^n + (b(a^n + b^n))^n = (a^n + b^n)^{n+1}.$$

In fact the $n + 1$ may be replaced by any k relatively prime to n, as Jaroslav has observed, since $(2^a)^n + (2^a)^n = (2^b)^k$ when $bk - an = 1$.

In a very short article in the *Monthly* for August–September 1988, Quine observed that if one has a row of bins, some of which are red, some blue, and some unpainted and if more than two objects are to be distributed into these bins, then Fermat's last theorem is equivalent to the assertion that the following statement is never true.

> The number of ways of sorting the objects so that both painted bins are shunned is equal to the number of ways that shun neither painted bin.

The argument is a simple bit of combinatorial reasoning and we leave it as an exercise.

We make one final observation. Chowla has shown that FLT is true if and only if the Diophantine equation

$$y^2 = 4x^p + 1$$

has no nontrivial rational solutions. (p is an odd prime.) A shortened version of this theorem has been given by Inkeri.

See also Integer 6, Sophie Germain's Theorem.

REFERENCES

[1] P. Ribenboim, Recent results on Fermat's last theorem, *Can. Math. Bull.* 54 (1977) 229–242.

[2] D. Shanks and H. C. Williams, Gunderson's function in Fermat's last theorem, *Math. Comp.* 36 (1981) 291–95.

[3] D. H. Lehmer, On Fermat's quotient, base two, *Math. Comp.* 36 (1981) 289–290.

[4] A. Granville and M. B. Monagan , The first case of Fermat's last theorem is true for all prime exponents up to 714 591 416 091 389, *Trans. Amer. Math. Soc.* 306 (1988) 329–359.

[5] D. Coppersmith, Fermat's last theorem (case 1) and the Wieferich criterion, *Math. Comp.* 54 (1990) 895–902.

[6] L. M. Adelman and D. R. Heath-Brown , The first case of Fermat's last theorem, *Invent. Math.* 79 (1985) 409–416.

[7] A. Granville, Powerful numbers and Fermat's last theorem, *C. R. Math. Rep. Acad. Sci. Canada* 8 (1986) 215–18.

[8] P. Ribenboim, Remarks on exponential congruences and powerful numbers, *J. Number Theory* 29 (1988) 251–263.

[9] J. H. Silverman, Wieferich's criterion and the abc-conjecture, *J. Number Theory* 30 (1988) 226–237.

[10] F. Hirzebruch, *Math. Rev.* 53 (1977) #11090.

[11] D. R. Heath-Brown, Fermat's last theorem for "almost all" exponents, *Bull. London Math. Soc.* 17 (1985) 15–16.

[12] W. C. Waterhouse, *Math. Rev.* 57 (1979) #12066ab.

[13] P. Ribenboim, *Thirteen Lectures on Fermat's Last Theorem,* Springer, New York, 1979.

[14] J. W. Tanner and S. S. Wagstaff , New congruences for the Bernoulli numbers, *Math. Comp.* 48 (198) 341–350.

[15] L. Hoehn, Solutions of $x^n + y^n = z^{n+1}$, *Math. Mag.* 62 (1989) 342.

[16] F. Jaroslav, *Math. Intelligencer*, 13 (1991) 33.

[17] W. V. Quine, Fermat's last theorem in combinatorial form, *Amer. Math. Monthly* 95 (1988) 636.

[18] S. Chowla, *Number Theory Day,* LNM #626 Springer, New York, 1977, pp. 1–43 (esp. 19–24).

[19] K. Inkeri, On certain equivalent statements for Fermat's last theorem—with requisite corrections, *Ann. Univ. Turku. Ser. A* 186 (1984) 12–22.

Integer 1621

7, 991, 1141, 1621, ...

To solve

$$x^2 = 1 + 7y^2$$

we may successively substitute the integers $1, 2, 3, \ldots$ for y in the right-hand side and examine the results until one finds a square. One finds $y = 3$, $x = 8$.

Question: Is the above method available if the 7 is replaced by 991?

Ask again after contemplating the following fact. Substitution of $1, 2, \ldots, 10^{28}$ for y will not yield a square. In fact, the smallest solution to this equation is given by:

$$x = 379\,516\,400\,906\,811\,930\,638\,014\,896\,080$$
$$y = 12\,055\,735\,790\,331\,359\,447\,442\,538\,767.$$

To emphasize the point we quote H. M. Stark as follows. (See the first reference below.)

> Perhaps the word "obvious" is beginning to lose its meaning, but to make sure, we give one last example. We see that $x = 1$, $y = 0$ satisfies the Diophantine equation

$$x^2 - 1141y^2 = 1.$$

We might ask, does the equation have any solutions in positive integers? We see that

$$x = \sqrt{1141y^2 + 1}.$$

Thus the question is: Is $1141y^2 + 1$ ever a perfect square? This may be checked experimentally. It turns out that the answer is no for all positive y less than 1 million. In view of the previous example, perhaps we should experiment further. The answer is still no for all y less than 1 trillion (1 million million, 10^{12}). We go overboard and check y up to 1 trillion trillion (10^{24}). Again, the answer is no. No one in his right mind would really believe that there could be a positive y such that $\sqrt{1141y^2 + 1}$ is an integer if there is no such y less than a trillion trillion. But there is. In fact, there are infinitely many of them, the smallest among them having 26 digits. If you still do not believe this, we will prove in Chapter 7 that there are infinitely many such y and give a method whereby you may start from scratch and find the smallest possible value of y in less than an hour (with a desk calculator).

An amusing such pair of equations is:

$$x^2 = 1 + 1620y^2 \qquad x^2 = 1 + 1621y^2.$$

The first of these has smallest solution with $x = 161$, $y = 4$ and the second has smallest solution with an x of 76 digits! (See the table in Whitford.) Is it mysterious that neighboring integers can act so very differently?

REFERENCES

[1] H. M. Stark, *An Introduction to Number Theory,* Markham, Chicago, 1970, pp. 7–8.

[2] W. Sierpiński, *Elementary Theory of Numbers,* North-Holland, Amsterdam, 1988, p. 94.

[3] R. D. Carmichael, *Diophantine Analysis,* Dover, New York, (reprint of 1915 edition), footnote, p. 33.

[4] E. Whitford, *The Pell Equation,* Coll. of the City of New York, New York, 1912.

Integer 1729

Hardy's Taxi

G. H. Hardy, speaking of Ramanujan, said,

> He could remember the idiosyncracies of numbers in an almost uncanny way. It was Littlewood who said that every positive integer was one of Ramanujan's personal friends. I remember going to see him once when he was lying ill in Putney. I had ridden in taxi-cab No. 1729, and remarked that the number seemed to me to be rather a dull one, and that I hoped that it was not an unfavourable omen. "No", he replied, "it is a very interesting number; it is the smallest number expressible as a sum of two cubes in two different ways." I asked him, naturally, whether he could tell me the solution of the corresponding problem for fourth powers; and he replied, after a moment's thought, that he knew no obvious example, and supposed that the first such number must be very large.

In fact the least integer representable as a sum of two fourth powers in two ways is the integer 635,318,657. Alter reports that this was known to Euler.

Alter denotes the least integer expressible as a sum of m positive nth powers in s ways by $r(m,n,s)$. In this notation, Ramanujan's observation says $r(2,3,2) = 1729$. What we said a moment ago asserts $r(2,4,2) = 635{,}318{,}657$ ($= 59^4 + 158^4 = 133^4 + 134^4$). It is not difficult to see that $r(2,2,2) = 50$ and $r(3,4,2) = 2673$. Further, Alter claims, for $4 \le t \le 15$, $r(t,4,2) = t + 255$, while, for $t \ge 16$, $r(t,4,2) = t + 240$. (Note that $t + 255 = 4^4 + 1^4 + \cdots + 1^4$ and $t + 240 = 3^4 + 3^4 + 3^4 + 1^4 + \cdots + 1^4$.)

Alter observes that it is not known if $r(2,5,2)$ or $r(2,6,2)$ exist.

REFERENCES

[1] G. H. Hardy, *Ramanujan: Twelve Lectures on Subjects Suggested by his Life and Work,* Chelsea, New York, (Reprint with 1940 Preface), p. 12.

[2] R. Alter, "Computations and generalizations on a remark of Ramanujan" in LNM #899, Springer, New York, 1981, pp. 183–196.

Integer 1794

The Binomial Coefficient $\binom{k^2}{k}$

Let p_i be the ith prime number. Then

$$p_1 \cdots p_k < \binom{k^2}{k} \qquad \text{for } 2 < k < 1794;$$

$$\binom{k^2}{k} < p_1 \cdots p_k \qquad \text{for } 1794 \leq k.$$

REFERENCES

H. Gupta and S. P. Khare, On $\binom{k^2}{k}$ and the product of the first k primes, *Publ. Fac. Êlectrotechn. Univ. Belgrade, Sér. Math. Phys.* 25–29 (1977) 577–598.

Integer 1848

Numeri Idonei

A well-known theorem in number theory states that an integer larger than 1 which is congruent to 1 modulo 4 is a prime if and only if it has a unique representation as a sum of relatively prime squares.

If the only odd integers greater than 1 which have unique representations in the form $x^2 + dy^2$, where x and y are relatively prime, are themselves prime, then one calls the integer d a *numerus idoneus.* The assertion above tells us that the integer 1 is such a number. Euler gave 65 instances of such numbers the largest of which was 1848.

In 1934 S. Chowla proved that the number of these numbers is finite. In the interim it has been shown that there is at most one such number which is square free and larger than 1365. (1365 is also a numerus idoneus.)

It is also known that non–square-free numeris idonei are either less than 100 or of the form $4d$ where d is a square-free even numerus idoneus.

Frei calls these numbers *convenient* numbers and has written an expository article about them (see below). He observes that they have connections with eigenvalue problems in physics.

REFERENCES

[1] W. Sierpiński, *Elementary Theory of Numbers,* North-Holland, Amsterdam, 1988, pp. 228–29.

[2] G. Frei, Leonhard Euler's convenient numbers, *Math. Intell.* 7 (1985) 55–58, 64.

Integer 2719

MR 56 #15557

S. Ramanujan [*Proc. Cambridge Phil. Soc.* 19 (1917), 11–21] listed the first 16 odd numbers (the last being 391) not representable as $x^2 + y^2 + 10z^2$ and stated that they did not seem to obey any simple law. The author [*Proc. Ind. Acad. Sci. Sect. A.* 13 (1941), 519–20; MR 3, 65] added 679 and 2719 to this list, so finding all such numbers below 20 000. This upper bound is now extended to 150 000. He conjectures that every odd number greater than 2719 can be expressed in the above form.

—R. A. Rankin (Glasgow)

REFERENCES

H. Gupta, Ramanujan's ternary quadratic form $x^2 + y^2 + 10z^2$, *Res. Bull. Panjab Univ.* 24 (1973) 57 (1977).

Integer 3304

Smith Numbers

A composite integer whose digit sum is the same as the sum of the digit sums of its prime factors is called a *Smith* number. The integer 4,937,775 is a Smith number as one can check from its prime factorization $3 \cdot 5 \cdot 5 \cdot 65387$.

Let $R_n = 11 \cdots 1$, where there are n 1's.

If R_n is prime and $n \geq 3$, then $3304R_n$ is a Smith number. Further, 3304 may not be replaced by any smaller integer and have this statement remain true.

Let $S(n)$ be the sum of the digits of n and let $S_p(n)$ be the sum of the digits in the prime factors of n.

1. If $S(n) > S_p(n)$ and $S(n) \equiv S_p(n) \pmod 7$, then $10^k n$ is a Smith number when $k = (S(n) - S_p(n))/7$.
2. If q is a prime with all digits either 0 or 1, then there is a c such that cq is a Smith number.

Wilansky tells us how these numbers were named in the following passage from the *Two Year College Mathematics Journal*

The number of Smith numbers between n thousand and n thousand + 999 for $n = 0, 1, \ldots, 9$, is, respectively, 47, 32, 42, 28, 33, 32, 32, 37, 37, 40.

I wonder whether there are infinitely many Smith numbers.

The largest Smith number known is due to my brother-in-law H. Smith who is not a mathematician. It is his telephone number: 4937775.

If $m = \prod_{i=1}^{k} p_i$, p_i prime, then McDaniel puts

$$S_p(b, m) = \sum_{i=1}^{k} S(b, p_i),$$

where $S(b, m)$ is the sum of the base b digits of m. If m is composite and c is defined by $S_p(b, m) - S(b, m) = c$, then we say that $m \in I_b(c)$. The elements of $I_{10}(0)$ are the Smith numbers.

McDaniel proves for c any integer in base b, $b \geq 8$, that $I_b(c)$ is infinite. Thus there **are** infinitely many Smith numbers.

REFERENCES

[1] S. Oltikar and K. Wayland, Construction of Smith numbers, *Math. Mag.* 56 (1983) 36–7.

[2] A. Wilansky, Smith numbers, *Two Year Coll. Math. J.*, 13 (1982) 21.

[3] W. L. McDaniel, Difference of the digital sums of an integer base b and its prime factors, *J. Number Theory* 31 (1989) 91–8.

Integer 4181

Converse Numbers

Let u_n be the nth Fibonacci number ($u_1 = u_2 = 1$) and $(\frac{a}{b})$ be the Legendre symbol.

The integer n is a *converse* number if

$$u_n \equiv (\frac{n}{5}) \pmod{n}.$$

All primes other than 5 are converse numbers and the smallest composite converse number is 4181.

REFERENCES

M. Yorinaga, On a congruential property of Fibonacci numbers— considerations and remarks, *Math. J. Okayama Univ.* 19 (1976) 11–17.

Integer 17163

Sums of Squares of Primes

The integer 17163 is the largest integer which is not the sum of distinct squares of prime numbers.

REFERENCES

R. E. Dressler, L. Pigno, and R. Young, Sums of squares of primes, *Nord. Mat. Tidskr.* 24 (1976) 39–40.

Integer 20161

Abundant Numbers

An *abundant* number is an integer which is smaller than twice the sum of all of its divisors.

Every integer larger that 20161 may be written as the sum of two abundant numbers while the same may be said for the even integers larger than 46.

In the review of the paper listed below it is stated that this result was anticipated by Parkin and Lander in 1964.

See Integer 77, Partitions.

REFERENCES

[1] S. Hitotumatu, On the limit for the representation by the sum of two abundant numbers, *Publ. Res. Inst. Math. Sci.* 8 (1972) 111–116.

[2] C. S.Ogilvy, J. T.Anderson, *Excursions in Number Theory,* Oxford Univ. Press, Oxford, 1966 (Dover 1988).

Integer 63001

Ramanujan's τ Function

The Ramanujan τ funtion is defined by the following equation.

$$x\prod_{m=1}^{\infty}(1-x^m)^{24} = \sum_{n=1}^{\infty}\tau(n)x^n = x - 24x^2 + 252x^3 + \cdots$$

Ramanujan stated three conjectures about this function all of which have now been proven. These conjectures were:

1. $\tau(nm) = \tau(n)\tau(m)$ for $(n,m) = 1$.
2. $\tau(p^{n+1}) = \tau(p)\tau(p^n) - p^{11}\tau(p^{n-1})$ for p a prime.
3. $|\tau(p)| \leq 2p^{11/2}$.

The third of these conjectures has only been proved in recent years and its proof is far from simple and was a consequence of the proof of the Weil conjectures by Deligne. A still unproved conjecture first stated by Serre and called the Sato–Tate conjecture reads as follows.

Conjecture 1 (Sato-Tate Conjecture). *Define θ_p by the equation*

$$\tau(p) = 2p^{11/2} \cos \theta_p,$$

where $0 \leq \theta_p \leq \pi$. Then the angles θ_p are uniformly distributed with respect to the measure $\frac{2}{\pi}(\sin \theta)^2 \, d\theta$.

It is known that $\tau(n)$ is composite for all n such that $2 \leq n \leq 63000$ but that $\tau(63001)$, which equals 80 561 663 527 802 406 257 321 747, is prime. ($63001 = 251^2$.)

It is known that given any positive integer m almost all values of $\tau(n)$ are divisible by m.

For further details and bibliography see the articles by Rankin, Murty, and Swinnerton-Dyer relating to the τ function in *Ramanujan Revisited.*

In a review of a paper by Zagier (83k:10056), Rankin makes the statement "Thus it appears that the values of the τ-function suffice to evaluate the zeros of the Riemann zeta-function!" The interested reader may consult either the review or the paper by Zagier cited below.

If we let p be a prime and a be an integer not divisible by p, then the number of n, not exceeding x, for which $\tau(n) \equiv a \pmod{p}$ is denoted by $\nu_a(x)$.

Radoux showed that there are constants α and C, independent of a, such that

$$\nu_a(x) \sim C \frac{x}{(\log x)^\alpha}.$$

I.e., the values of $\tau(n)$ are equidistributed among the $p - 1$ nonzero residue classes modulo p.

Niebur gave the following expression for $\tau(n)$.

$$\tau(n) = n^4 \sigma(n) - 24 \sum_{k=1}^{n-1} (35k^4 - 52k^3 n + 18k^2 n^2) \sigma(k) \sigma(n-k), \quad n \geq 1$$

and used it to search for primes p such that $\tau(p) \equiv 0 \pmod{p}$.

The only primes ≤ 65063 found were 2, 3, 5, 7, 2411. This extends a previous search carried out by M. Newman.

See also Integer 23, The τ Function, and Integer 691, A Divisibility Property.

REFERENCES

[1] D. H. Lehmer, The primality of Ramanujan's τ function, *Amer. Math. Monthly* Slaught Mem. Papers 10 (1965) 15–18.

[2] G. E. Andrews, et al. (eds.), *Ramanujan Revisited,* Academic Press, Boston, 1988.

[3] P. Deligne, La conjecture de Weil I, *Inst. Hautes Ètudes Sci. Publ. Math.* 43 (1974) 273–307.

[4] D. Zagier, The Rankin–Selberg method for automorphic functions which are not of rapid decay, *J. Fac. Sci. Univ. Tokyo Sect. 1A Math.* 28 (1981) 415–437 (1982).

[5] C. Radoux, Répartition des valeurs de la fonction τ de Ramanujan modulo un nombre premier, *Ann. Soc. Sci. Bruxelles Sér. I* 89 (1975) 434–38.

[6] D. Niebur, A formula for Ramanujan's τ-function, *Ill. J. Math.* 19 (1975) 448–49.

Integer 720720

A Divisibility Problem

The number 720720 is the largest integer which is divisible by all of the integers not exceeding its fifth root.

In Ozeki similar results are also proved for the kth roots for $6 \leq k \leq 10$. Results for $k = 2, 3, 4$ were known previously.

For the case $k = 2$ see Integer 24, A Property of 24 .

REFERENCES

N. Ozeki, On the problem $1, 2, 3, \ldots, [n^{1/k}] | n$, *J. Coll. Arts. Sci. Chiba Univ.* 3 (1961/62) 427–431.

Integer 99925853

For each positive integer k, Erdös lets S_k stand for the assertion that there exist coprime integers a, b with $b - a = k$ and such that $(ab, c) > 1$ for all c strictly between a and b.

For k even, we see that S_k is false by taking $c = \frac{a+b}{2}$. Further, if S_d fails, then so also does S_{td} for all positive integers t. The smallest known k for which S_k is known to hold is the prime 99925853. It is not known if S_k is true for infinitely many k.

REFERENCES

P. Erdős, "Some problems in number theory" in *Graph Theory and Computing,* 17th Boca Raton Conference on Combinatorics, Boca Raton, 1986.

Integer 906180359

Liouville's function $\lambda(n)$ is defined to be $(-1)^{\nu}$, where ν is the number of (not necessarily distinct) prime factors of n. For example $\nu(18) = 3$. We put $L(x) = \sum_{n \leq x} \lambda(n)$. In 1960 R. S. Lehman showed that $L(906180359) = 1$, confirming the falsity, known since the work of Haselgrove in 1958, of a conjecture of Pólya. In the paper by Tanaka, this is shown to be the very first value of $x > 1$ for which $L(x) > 0$.

Chowla observes that the Riemann hypothesis is equivalent to the assertion of the equality

$$L(x) = O\left(x^{(1/2)+\epsilon}\right),$$

where ϵ is any positive quantity.

If one writes out the first 16 terms of the sequence $\{\lambda(n)\}$, one obtains

$$1, -1, -1, 1, -1, 1, -1, -1, 1, 1, -1, -1, -1, 1, 1, 1.$$

It will be noted that all of the eight possible sequences of three 1's and -1's appear somewhere in this sequence of 16 terms.

Hildebrand has shown that in the full infinite sequence every one of the eight possible triples appears infinitely many times.

If one denotes by L, respectively M, the set of positive integers with the property that for each prime number p dividing the integer so also does the number p^2, respectively p^4, divide the integer, then the following amusing identities obtain.

$$\sum_{n\in L} \frac{\lambda(n)}{\phi(n)} = \zeta(2) = \frac{\pi^2}{6} \qquad \sum_{n\in M} \frac{\lambda(n)}{\phi(n)} = \frac{\zeta(4)\zeta(6)}{\zeta(12)}$$

There is a somewhat related problem having to do with the prime factors of positive integers. It involves the Möbius function μ—defined by $\mu(1) = 1, \mu(n) = 0$, if $n > 1$ and is divisible by a square larger than 1, and $\mu(n)$ is 1 (or -1) if n is a square-free product of an even (or odd) number of primes.

The number of square-free integers not exceeding x which have an even number of prime factors diminished by the number having an odd number of prime factors is given by $M(x)$ where

$$M(x) = \sum_{n\leq x} \mu(n).$$

In 1885 Stieltjes told Hermite in a letter that he could prove that $\frac{M(x)}{\sqrt{x}}$ is bounded and that probably ± 1 would serve as bounds. His "proof" was never published. The conjecture was again formulated in a paper by Mertens in which he gave a 50 page table of values of $\mu(n)$ and $M(x)$ for values up to $x = 10000$. In recent years the conjecture has been known as the Mertens' conjecture.

Between 1897 and 1913 von Sterneck extended Mertens' data and conjectured that

$$M(x) < \frac{\sqrt{x}}{2}, \qquad \text{for } x \geq 200.$$

In 1960 Jurkat disproved this conjecture without giving an explicit counterexample. More recently, Cohen and Dress have shown that the least x for which the inequality is reversed is $x = 7725038629$ with the value 43947 for $M(x)$.

In 1985 Odlyzko and te Riele disproved Mertens' conjecture—again without giving an explicit counterexample. See their paper for a history of the problem and a list of 45 references. The disproof of this conjecture afforded one of those rare instances where a result in mathematics received a prominent place in the popular press.

REFERENCES

[1] C. B. Haselgrove, A disproof of a conjecture of Pólya, *Mathematika* 5 (1958) 141–45.

[2] R. S. Lehman, On Liouville's function, *Math. Comp.* 14 (1960) 311–320.

[3] M. Tanaka, A numerical investigation on cumulative sum of the Liouville function, *Tokyo J. Math.* 3 (1980) 187–89.

[4] S. Chowla, *The Riemann Hypothesis and Hilbert's Tenth Problem,* Gordon and Breach, New York, 1965.

[5] A. Hildebrand, On consecutive values of the Liouville function, *Ens. Math.* 32 (1986) 219–226.

[6] D. Suryanarayana, Note on Liouville's λ-function and Euler's ϕ-functions, *Math. Stu.* 37 (1969) 217–18.

[7] A. M. Odlyzko and H. J. J. te Riele , Disproof of the Mertens conjecture, *J. für die Reine und Angew. Math.* 357 (1985) 138–160.

Integer 10513222720

A Method of Factoring

Behold! From 10513222720=1020305 · 10304 one determines a factorization of

$$x^5 - 5x^4 + 13x^3 - 22x^2 + 27x - 20$$

to be the product of $x^3 - 2x^2 + 3x - 5$ and $x^2 - 3x + 4$.

This does look like a one of a kind factorization but it is the result of applying a systematic method of factoring to the fifth degree polynomial given. For details see the references below.

REFERENCES

M. V. Yakovkin, On a method of finding irreducible factors, *Dokl. Akad. Nauk SSSR* (N.S.) 93 (1953) 783–85.

——, *Numerical Theory of Polynomial Reduction* (in Russian), Akad. Nauk, CCP, 1959.

Integer 608981813029

A Chebyshev Observation

Let $\pi_{b,c}(x)$ be the number of primes not exceeding x which are congruent to c modulo b.

Bays and Hudson report the following. "In a letter written in 1853 Chebyshev ... remarked that

$$\pi_{3,1}(x) < \pi_{3,2}(x)$$

and

$$\pi_{4,1}(x) < \pi_{4,3}(x)$$

for all small values of x."

(In a paper, cited below, Hudson and Caron offer a "combinatorial explanation" of why $\pi_{4,3}(x) - \pi_{4,1}(x)$ is approximated by $\frac{\sqrt{x}}{2}$ for all "small" x.)

By a 1914 result of Littlewood, one knows that not only are these inequalities true for an infinity of x but also the reverse inequalities are true for an infinity of x. Thus the question arises as to where the inequality first reverses.

The first negative value for $\pi_{4,3}(x) - \pi_{4,1}(x)$ was found to be 26861 by Leech in 1957. The result was found by computer.

A similar investigation by Bays and Hudson for the quantity $\pi_{3,2}(x) - \pi_{3,1}(x)$ turned out to require considerably more computer power. Nevertheless, they "discovered on December 25, 1976" that the first negative value occurred for $x = 608{,}981{,}813{,}029$.

Letting the above value of x be denoted by x_0 and putting $x_f = 610{,}968{,}213{,}796$ one finds 316,889,212 values of x between x_0 and x_f for which the quantity is negative and these are split into two groups consisting of 150,276,170 and 166,613,042 numbers, respectively, and the groups are themselves separated by 1,363,263,116 integers.

Littlewood's theorem tells us that if

$$F(x) = \frac{\pi(x) - Li(x)}{\frac{x^{1/2}}{\log x} \log\log\log x},$$

then $\limsup_x F(x) \geq \frac{1}{2}$ and $\liminf_x F(x) \leq -\frac{1}{2}$. (Here $\pi(x)$ is the number of primes not exceeding x and $Li(x)$ is the logarithmic integral of x.)

From this it is clear that infinitely often $\pi(x)$ is greater than $Li(x)$ and infinitely often it is less than $Li(x)$.

Elliott observes that Littlewood's theorem is equivalent to the law of the iterated logarithm in probability theory. He also links the difference in the two quantities $\pi(x)$, $Li(x)$ to the central limit theorem and to the changes in the lead in sequences of coin tossing results.

In 1987 te Riele showed that there were more than 10^{180} successive integers in the range from $6.62 \cdot 10^{370}$ to $6.69 \cdot 10^{370}$ for which $\pi(x) - L(x) > 0$. Thus, though $\pi(x) < L(x)$ for all small values of x (e.g., all x smaller than a billion) the inequality reverses itself somewhere before $6.69 \cdot 10^{370}$. This is considerably better than the number $10^{10^{10^{34}}}$ proved by Skewes in 1955 to be greater than the first reversal point for the inequality.

REFERENCES

[1] C. Bays and R. H. Hudson, Details of the first region of integers with $\pi_{3,2}(x) < \pi_{3,1}(x)$, *Math. Comp.* 32 (1978) 571–76.

[2] R. H. Hudson and A. Caron, A common combinatorial principle underlies Riemann's formula, the Chebyshev phenomenon, and other subtle effects in comparative prime number theory I, *J. für die Reine und Angew. Math.* 313 (1980) 133–150.

[3] P. D. T. A. Elliott, The Riemann zeta function and coin tossing, *J. für Reine und Angew. Math.* 254 (1972) 100–109.

[4] H. J. J. te Riele, On the sign of the difference $\pi(x) - li(x)$, *Math. Comp.* 48 (1987) 323-28.

Integer 2935363331541925531

Two Sequences of Composite Integers

For $n = 2935363331541925531$ every number in the sequence

$$1 + 2n, 1 + 4n, 1 + 8n, 1 + 16n, \ldots$$

is composite.

Selfridge has replaced this n by the number 78557.

Not all terms in the Fibonacci sequence are composite as is immediate upon looking at the first few terms. However, there are infinitely many composite terms since every term F_{3k} is divisible by 2.

In 1964 Graham showed how to find relatively prime integers a, b such that the sequence $\{A_n\}$, where

$$A_0 = a, \quad A_1 = b, \quad A_n = A_{n-1} + A_{n-2}, \qquad \text{for } n \geq 2,$$

has only composite terms. The calculations led to a and b each having 34 digits.

Recently Knuth has corrected a numerical error in Graham's paper and at the same time modified the algorithm to yield the following 17 digit values for a and b.

$$a = 62638280004239857 \qquad b = 49463435743205655$$

Even more recently Wilf has, in a letter to the editor of the *Mathematics Magazine,* given an a, b pair slightly smaller than these. Basing his work on Graham's sequence, Wall has argued that if g_n is the nth term in this sequence and if $x_n = 2^{12} \cdot 8209 \cdot g_n$, then the sequence of x_n is a Fibonacci-like sequence of abundant numbers. We leave determining whether the numerical error Knuth pointed out in Graham's paper vitiates this result to the reader.

It is not known if the Fibonacci sequence contains infinitely many primes but it is easy to show, however, that if u_n is prime, then n itself must be prime. (This follows from the fact that u_k divides u_{km}. As an aside, we note that u_{mn} is divisible by $u_m u_n$ if and only if the greatest common divisor of m and n is 1, 2, or 5. See Jarden, cited below.) Unfortunately, there are prime n for which u_n is composite as one sees from the example $u_{19} = 37 \cdot 113$.

In view of this last it might appear especially surprising that the following sequence $\{U_n(x)\}$ of "Fibonacci polynomials" has the property that $U_n(x)$ is irreducible if and only if n is prime.

$$\begin{aligned} U_0(x) &= 0, \qquad U_1(x) = 1, \\ U_n(x) &= xU_{n-1}(x) + U_{n-2}(x) <?? \qquad \text{for } n \geq 2. \end{aligned}$$

This is due to Parberry and Webb.

$$\frac{y}{1 - xy - y^2} = \sum_{n=0}^{\infty} U_n(x) y^n$$

is a generating function for the U_n.

This sequence of polynomials starts out

$$0,\ 1,\ x,\ x^2 + 1,\ x^3 + 2x,\ x^4 + 3x^2 + 1,\ \ldots,$$

and it is easy to see that $U_n(1) = u_n$. Neglecting the first term and taking $x = 1$ we find the usual generating function

$$\frac{1}{1 - y - y^2}$$

for the Fibonacci numbers. (Golomb has found

$$\frac{1 - x}{1 - 2x - 2x^2 + x^3}$$

as a generating function for the squares of the Fibonacci numbers and this has been extended to the kth powers by Riordan.)

It is well known that every prime divides some Fibonacci number (see Hardy and Wright p. 150) and that consecutive Fibonacci numbers are always relatively prime. Thus always, for $n > 1$, u_{n+1} contains a prime factor not dividing u_n and at some point in the sequence every prime appears. If we let u_α denote the first Fibonacci number divisible by the prime p, then α $(= \alpha(p))$ is called the *rank of apparition* of p. Jarden discusses this and shows that α divides $p - (\frac{5}{p})$ and, indeed, the quotient is an unbounded function of p.

In Jarden's argument he proves a lemma of some independent interest.

Lemma. *Let $k_1, \ldots, k_s$ be an arbitrary set of s positive integers. Then there is a prime $q > 5$ such that all of the numbers*

$$k_1q \pm 1,\ k_2q \pm 1, \ldots,\ k_sq \pm 1$$

are composite.

The argument is a simple application of Dirichlet's theorem on arithmetic progressions. Put $A = \prod_{j=1}^{s}(3k_j - 1)(3k_j + 1)$ and note the existence of a prime q, say for $x = x_q$, in the sequence

$$3 + xA, \qquad x = 0, 1, 2, \ldots.$$

Then $k_i q \pm 1 = k_i(3 + x_q A) \pm 1$ is divisible by the number $3k_i \pm 1$, which is greater than 1 and smaller than $k_i q \pm 1$ itself. Since this is true for each i the argument is complete.

One might ask how often a prime occurs in u_n which has not occurred in any of the earlier u_k. As has been observed in Integer 12, Prime Factors of Fibonacci Numbers, it is the norm since, as Carmichael showed in 1913, only for $n = 1, 2, 6, 12$ does u_n **not** contain some prime not dividing any earlier Fibonacci number.

Shallit and Yamron give an interesting third order recurrence $\{T_n\}$ with the property that when p is a prime then $p|T_p$. In fact, put $T_0 = 3$, $T_1 = 0$, $T_2 = 2$, $T_k = T_{k-2} + T_{k-3}$, for $k \geq 3$.

Though p prime is sufficient for T_p to be divisible by p it is not necessary since, as one may check, $521^2|T_{521^2}$. The only other composite integer less than a million for which this is true is

$$n = 904631 = 7 \cdot 13 \cdot 9941.$$

Asymptotically T_k is equal to α^k, where α is the real root of the equation $x^3 - x - 1 = 0$.

See also Integer 12, Prime Factorizations of Fibonacci Numbers.

REFERENCES

[1] H. M. Stark, *An Introduction to Number Theory*, Markham, 1970, Problem 14, pp. 110–111.

[2] R. L. Graham, A Fibonacci-like sequence of composite numbers, *Math. Mag.* 37 (1964) 322–24.

[3] D. E. Knuth, A Fibonacci-like sequence of composite numbers, *Math. Mag.* 63 (1990) 21–25.

[4] H. S. Wilf, Letter to the Editor, *Math. Mag.* 63 (1990) 284.

[5] C. R. Wall, A Fibonacci-like sequence of abundant numbers, *Fib. Quarterly* 22 (1984) 349.

[6] G. H. Hardy and E. M. Wright, *An Introduction to the Theory of Numbers* (5th ed.), Oxford Univ. Press, Oxford, 1988.

[7] D. Jarden, Two theorems on Fibonacci's sequence, *Amer. Math. Monthly* 53 (1946) 425–27.

[8] W. A. Webb and E. A. Parberry , Divisibility properties of Fibonacci polynomials, *Fib. Quarterly* 7 (1969) 457–463.

[9] J. Riordan, Generating functions for powers of Fibonacci numbers, *Duke J.* 29 (1962) 5–12.

[10] J. G. Shallit and J. P. Yamron , On linear recurrences and divisibility by primes, *Fib. Quarterly* 22 (1984) 366–68.

Integer 112359550561797752809

A Multiple of 99

If one places a digit 1 at the left and at the right of the given integer, one obtains an integer which is 99 times the original integer and no smaller integer has this property.

Hunsucker and Pomerance generalize the question as follows. In base g find integers x, k such that

$$g^{k+2} + gx + 1 = (g^2 - 1)x, \ k = [log_g x].$$

The authors claim that if one chooses a base g "randomly," then the "probability" of this generalized problem not having a solution is 1.

REFERENCES

J. L. Hunsucker and C. Pomerance, On an interesting property of 112359550561797752809, *Fib. Quarterly* 13 (1975) 331–33.

Integer 357686312646216567629137

Truncatable Primes

The integers 3137, 313, 31, 3 are all primes so we call the integer 3137 *right truncatable.* Since 3137, 137, 37, 7 are also all prime we say that 3137 is *left truncatable*.

The largest right truncatable prime is 73939133 and the largest left truncatable prime is 357686312646216567629137.

In base n the largest and the least truncatable primes are given in the table below.

n	R	L
3	71	23
4	191	4091
5	2437	7817

For base 10, Walstrom and Berg give the following table of numbers which yield, through right truncation, the entire set of right truncatable primes.

53	317	599	797	2393	3793	3797
7331	23333	23339	31193	31379	37397	73331
373393	593993	719333	739397	739399	2399333	7393931
7393933	23399339	29399999	37337999	59393339	73939133	

Chawla, Maxfield, and Muwafi consider the same problem with the exception that they regard 1 as a prime. Also, instead of right (left) truncatable primes they refer to left- (right-) handed primes. They argue that there are just 147 left-handed primes, give a table of all of them, note that the largest is 1979339339, list all right-handed primes up to 12953, and conjecture there are infinitely many of them.

One might ask for primes such that every permutation of their digits yields a prime. Richert calls such primes *permutation primes* and shows that other than primes with all digits 1 there are no permutation primes with n digits for any n satisfying the inequality $3 < n < 6 \cdot 10^{175}$.

The only three digit permutation primes are 113, 199, 337 and their permutations.

A fairly extensive table of base 10 truncatable primes is given in the paper by Atanassov cited below. In that paper the author also considers related sorts of numbers; for example, those that are prime and stay prime under successive removals of digits simultaneously from the right and left.

Pinch states that the largest right truncatable primes in bases $2, 3, \ldots, 9$ are, respectively, 1011, 2122, 2333 or 133313, 34222, 2155555, 25642 or 166426, 2117717, 3444224222.

REFERENCES

[1] I. O. Angell and H. J. Godwin, On truncatable primes, *Math. Comp.* 31 (1977) 265–67.

[2] J. E. Walstrom and M. Berg, Prime primes, *Math. Mag.* 42 (1969) 232.

[3] L. M. Chawla, J. E.Maxfield, and A. Muwafi, On the left-handed, right-handed and two-sided primes, *J. Nat. Sci. and Math.* 7 (1967) 95–99 (Correction 7 (1967) 259).

[4] H.-E. Richert, On permutation prime numbers, *Norsk. Mat. Tidsskr.* 33 (1951) 50–53.

[5] K. T. Atanassov, The most prime prime numbers, *Bull. No. Thy. and Related Topics* 9 (1985) 4–6.

[6] R. Pinch, Unsolved problem **ASI 88:13** in *Number Theory and Applications* (ed. R. A. Mollin), Kluwer, Dordrecht, 1988, p. 597.

WORD/PHRASE INDEX

INTEGER INDEX